MAKE YOUR SUMMER GARDEN LAST ALL YEAR

Patricia Shannon Kulla

Drawings by
Marilyn Miller

Library of Congress Catalog Card Number: 75-8460
ISBN 0-915336-01-4

For the Meyer girls—
Rose, Amelia, Mina, Ann and Martha—
fresh food mavins all.

TABLE OF CONTENTS

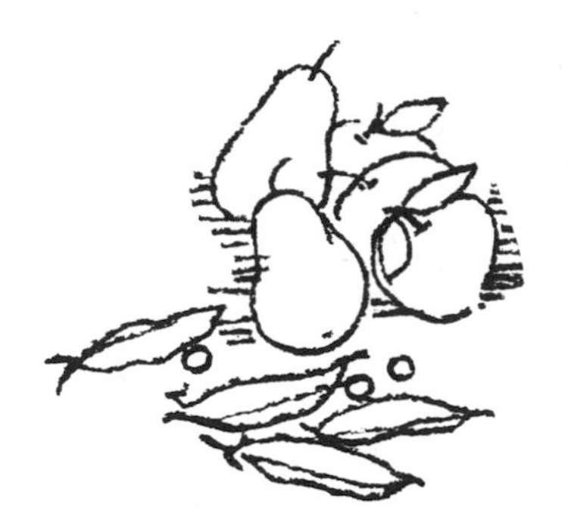

INTRODUCTION

This book is a product of frustrations—not a single frustration, mind you, but a whole collection of them. To begin, there was the frustration with food prices that never seem to stop rising and food quality that never seems to stop diminishing. There was not a great deal to be done about some things, of course. Becoming a cattle rancher, dairy, pig, poultry or grain farmer was out of the question. But fresh produce—vegetables, fruits, herbs—was another matter. Surely there was a way to combat the frustration of buying tomatoes, for example, in all but the height of the summer growing season, that have the taste and texture of styrofoam, and probably the nutritional value as well.

A summer garden was the solution. We'd planted them before. In fact, I come from a family of dedicated gardeners who begin poring over seed catalogues and planning for the next harvest not more than a week or two after the last harvest ends. The garden produced and for a few months we had mouthwatering, fresh-picked, vine-ripened produce. But three months, even stretching it, four months out of twelve is a short time. Couldn't we somehow make the joys of the garden last longer?

We froze what we could, but freezer capacity is limited as is the time a lot of frozen produce will hold in good condition. (There is also the rising energy cost to operate freezers, and the concern over the effect of any prolonged power failures.) We did canning, too, and the basement storage shelves were soon filled with jars of vegetables, fruits, pickles, jams and jellies. But once you've canned thirty quarts of pears, what do you do with the other two bushels from a neighbor's bountiful tree? And once you've put up catsup, chili sauce, an assortment of relishes and fifty quarts of tomatoes, what do you do with the rest still ripening on the vine? Satisfying as home-canned or frozen produce may be, what can be done to bring natural, fresh produce to the table the year round?

In my search for the answers, I quickly discovered that they could not be found in any single source. A comment or reference in one book led to further comments and references in others. Letters to agricultural colleges, extension services and government agencies brought bits of information here and there, along with referrals to other colleges, services and agencies. I felt a little like Robert Benchley who innocently looked up a footnote reference in a book late one evening and was discovered in his library the following morning literally inundated with books, moving sleepy-eyed from footnote to footnote.

What started as a personal research effort soon evolved into a full-scale book research project. If the kind of information I was looking for had never been gathered together into a single volume before, it was time that it should be. We began by investigating procedures for storing fresh produce. This led into drying techniques, since much of the produce we use is preserved and stored in this way—peas, beans, herbs, raisins, etc. Finally, we looked into prospects for year-round growing, we've called it Winter Growing, and found a surprisingly wide range of interesting possibilities.

The end result, this book, is far from a solo effort. There were a great many people involved; there are a great many people to thank. There are all the specialists and experts at the agricultural colleges, extension services and government agencies I've mentioned previously. There are friends and relatives, nearly all of whom had favorite methods that they willingly contributed. Special thanks are due Mitchell Brown, who helped gather the original research, and Dean Gottehrer, who helped organize and write many of the individual articles herein.

From a collection of research and reference materials weighing nearly forty pounds, we have compiled a book weighing less than a pound. All the essential information is here. If you follow it, I think you'll find, as I did, that you can indeed "make your summer garden last all year."

Patricia Shannon Kulla
Wilton, Connecticut
April, 1975

DRYING

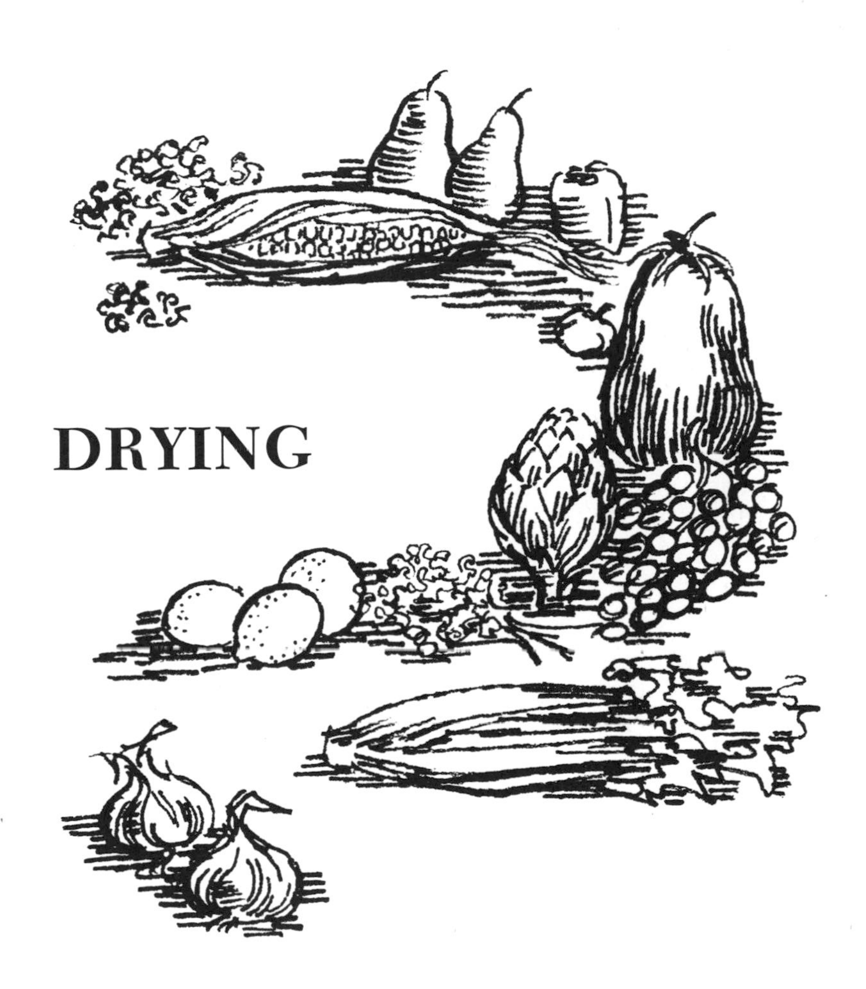

Drying is one of the oldest methods of food preservation known to man. Prehistoric man learned by experience that certain grains, peas, and beans left on the vine would dry as they matured and could be stored over a period of time to be used later when none remained to harvest.

In the Egyptian cradle of civilization and in other tropical and temperate climates, figs and dates grew in abundance. These fruits were natural candidates for drying because of their low moisture and high sugar content. From earliest times, they were dried along with grapes by being buried in the hot desert sand. Today, in some instances, they are still dried in much the same manner.

Dried figs were known to the ancient Greeks. In the 4th century B.C., the poet Alexis of Thurii related how the diet of an impoverished family included "that god-given inheritance of our mother country, darling of my heart, a dried fig." Ancient oriental writings, scrolls, and hieroglyphics also mention the drying of dates, raisins, and figs.

In the Middle Ages, European cooks used dried beans, peas, and whole grains in their meals. When Columbus discovered America he had been searching for a new route to bring the rare, exotic dried herbs and spices to Spain from the Far East. Explorers following in Columbus' wake carried dried peas to provide variety for the crew from the salt beef and pork that were then the staples of a sailor's diet.

In Latin America at the time of the conquistadors, lowland Peruvians ground dried maize kernels into flour and boiled it into a porridge. In the highlands above 11,000 feet where maize would not grow, the Peruvians used a combination of freezing and drying to preserve potatoes. After the harvest, the potatoes were spread on the ground and left overnight in the cold mountain air. The next day, men, women, and children stomped on the potatoes with their feet in much the same way as Europeans mashed their grapes to make wine. This procedure was repeated for four or five days until the potatoes had lost most of their moisture. They were left to finish drying in the sun and then stored. *Chuño,* as the dried potato was known to the Peruvians, was used to feed the masses, especially the slave workers in the silver mines.

After Columbus, early American settlers and frontiersmen learned the techniques of drying from the Indians. In pioneer times most farmhouses had an outdoor stone oven separate from the house. The oven door was about three feet wide and the oven itself as much as two feet high in the center and six feet deep. Heated by a stack of wood, the oven was used by the farm housewife to bake breads, cookies, cakes, and pies. In addition, the fresh fruits of summer, cultivated or wild, were dried in the oven. Apples, peaches, pears, plums, cherries, and all kinds of berries were slowly dried, and then stored in the attic until needed in the winter.

The dried food industry had its informal beginnings in the United States when the priests of the Spanish missions planted orchards along the Pacific coast. Trays of raisins, prunes, and other fruits were dried in the warm California sun and air, which was perfect for this purpose.

The technology of the dehydration process was not perfected until 1795 in France where hot air dehydration chambers were used. Canning was perfected a few years later, also in France. And canning eventually became popular in the U.S. as well. During World War II, however, drying was more in vogue because sugar and canning equipment were scarce. Similar considerations, a concern for natural foods, and the rising popularity of camping and backpacking, have brought about a renewed interest in dried foods and drying techniques today.

Advantages and Disadvantages of Drying

Drying is one of the simplest, most economical and ecologically sound methods of produce preservation. The only expenses for sun-dried foods, other than the produce itself, are for the trays and other drying equipment which can either be bought or built. No electricity or gas is needed to sun dry, no sugar is used as a preservative, no freezer or refrigerator is required for storage. Controlled heat drying requires a minimal amount of gas or electricity to heat an oven at a low temperature.

Dried foods are concentrated, compact, and require no special storage facilities. Home-dried foods are less expensive than store-bought dried foods and are excellent provisions for camping, hiking, or cross-country biking. Dried foods can be easily stored in plastic bags shaped to fit almost any tiny space in a backpack or camping bag.

The major disadvantage to preserving foods by drying is the amount of time it takes—as much as a week to ten days. Yet most of the labor is done by the sun or an oven. Your labor contribution is to select, prepare, and pretreat the food, then check to see when it is dry.

One other factor, neither an advantage nor disadvantage, is that dried foods have a different appearance and taste from the same food in its fresh, raw state. It is almost impossible to restore food preserved by any method to the same appearance and taste the food had before preservation. Dried foods taste different from the same fresh foods, have a different consistency, and different nutritional values. This is not to say that dried foods are superior or inferior, just that they are different. Some foods can be eaten in the dried state—notably pineapple, apricots, peaches and other fruits. Other dried foods, most vegetables for example, are best cooked before they are eaten. Home-dried foods taste different from store-bought dried foods and here again the taste and appearance are not necessarily superior or inferior, just different.

Not all foods lend themselves to drying. Some fruits and vegetables are more readily dried than others. Generally, fruits are more easily dried than vegetables, since fruits have higher concentrations of sugar and acid than vegetables. Some fruits—strawberries, raspberries and blueberries, for example—are better canned or frozen, although they can be dried.

According to the United States Department of Agriculture, the following fruits lend themselves most readily to home drying: apples, apricots, cherries, coconuts, dates, figs, guavas, nectarines, peaches, pears, plums, prunes. Among the fruits less readily dried are: avocados, blackberries, bananas, breadfruit, dewberries, loganberries, mamey (a tropical fruit), grapes.

Again, according to the U.S. Department of Agriculture, the vegetables most readily home dried are: beans (in the mature state: kidney, lima, mongo, pinto, pole, red, black, soy), beans (in the green state: lentils, soy), chili peppers, herbs, mature peas (sugar peas, cow peas, chick peas, pigeon peas), sweet corn, sweet potatoes, onions, soup mixture. The vegetables less readily dried are: asparagus, beets, broccoli, carrots, celery, greens (collards, mustards, turnip tops, beet tops, sweet potato leaves), green snap beans, green peas, okra, green peppers, pimientos, pumpkin, squash, tomatoes.

The list merely gives an idea of the relative ease with which the mentioned fruits and vegetables can be dried. *All* of them can be dried, of course—and they are not the only ones. A complete and detailed directory appears at the end of this section.

What Drying Does to Produce

One of the basic objectives of food preservation is to prevent decay and spoilage that will occur if food is not properly treated. Microorganisms—molds, yeasts, and bacteria—are the major cause of fruit and vegetable spoilage. Since all are living cells that require moisture to grow, drying combats their action by lowering the moisture content in the food.

Molds are sometimes highly prized; they are the elements responsible for the veins in blue cheeses, for example. On other foods, they spoil the taste and smell and make them generally unappetizing. Yeasts cause fermentation which is enjoyable in beer and necessary for sauerkraut, but unappealing on fresh fruit. Some bacteria—botulinus, salmonella, and staph—produce toxins or cause diseases that can be fatal.

Molds usually grow on foods containing above 12 percent moisture but have been known to grow on foods containing as little as 5 percent. Both yeasts and bacteria usually require a moisture content above 30 percent to flourish.

Dried fruits generally contain a higher percentage of moisture than dried vegetables. The moisture level of most dried fruit ranges between 16 and 25 percent, eliminating the possibility of spoilage due to yeast and bacteria. Sealing the fruit in an airtight container and storing it in a dry place eliminates the possibility of mold, even though the moisture level is still in the range where mold can grow.

Another cause of food spoilage is enzyme activity, a natural process at work in all organic matter. Left untreated, enzyme action causes fruits and vegetables to decompose, resulting in changes of color, texture, and taste that make the produce unpalatable. Enzyme activity is most pronounced at temperatures ranging between 85°F and 120°F and begins to slow around 140°F. During the drying process, enzyme activity can be checked by treating food with moist heat (steaming or blanching) at temperatures above 140°F for a specified period of time.

The process of drying foods changes their chemical

and physical traits, their color and appearance. These changes are most evident in fruits, which are rich in carbohydrates that may be discolored by enzyme action and the higher temperatures used to halt it. Some fruits are especially prone to discoloration when they come in contact with air, especially during slow sun drying. The discoloration of fruit can be halted by using chemicals such as sulfur, salt, or ascorbic acid (vitamin C). Dried fruits sold in supermarkets usually have been treated with sulfur to retain their natural colors. Unsulfured dried fruits sold in health food stores are generally darker than the same dried fruits sold in supermarkets.

Drying can also change the nutritional value in food. By eliminating moisture and reducing the bulk, natural sugars in dried fruits are concentrated and their caloric content increases. Nearly all dried produce has higher protein, fat, and carbohydrate levels because moisture and bulk have been reduced.

Drying preserves some vitamins and destroys others. Vitamin A is destroyed in sun dried foods by the sun's rays; it is retained for a while in foods dried by controlled heat, but gradually disappears.

The preservation of B vitamins—thiamine, riboflavin and niacin—depends on pretreatment and storage methods. Thiamine is destroyed by heat or sulfuring. Riboflavin is preserved as long as the food is stored in a completely dark place. If the food is exposed to light, riboflavin is destroyed. Niacin appears to be unaffected by drying or storage procedures.

Sulfuring destroys thiamine, but helps to preserve vitamin C which diminishes over time without sulfur treatment. Sulfuring is a personal choice—nutritional and other arguments pro and con are discussed in the instructions for sulfuring later in this section.

The methods used to prepare dried or fresh foods for consumption affect their nutritional value. All dried foods containing B-complex vitamins should be served with the water in which they were cooked or rehydrated. Since B vitamins are water soluble, cooking leaches the vitamins into the water. Tossing out the cooking water tosses out some of the food's natural vitamin content.

A judicious haste while harvesting, preparing, and drying foods makes a better product. Enzyme changes begin rapidly once food is harvested, so the less time that elapses between harvesting and drying, the better the product will appear and taste after it is dry. For the best flavor, color, texture, and nutritional value all dried foods should be used within one year.

Drying: A Quick Overview

An abbreviated summary will help you understand the entire drying process from start to finish before we go on to explain it in detail.

- Select the fruits or vegetables to be dried. Only the best quality will do because drying does not improve foods. Until the food is placed in the dryer, speed is essential from harvest to dryer. The longer it takes to process any food, the more it will deteriorate.
- Wash the food if necessary, then peel, slice, core, pit, etc., to prepare the food for drying. Fruits require temporary treatment as soon as they are cut to keep them from discoloring.

● To halt enzyme and prevent microorganic action, foods must be pretreated before drying. Fruits must be sulfured or treated in some alternative way; vegetables must be steamed.

● Place the food on drying trays and put it in the sun or in a controlled heat dryer.

● When the food is dry, remove it from the dryer and condition it for several days to two weeks in containers to equalize the remaining moisture and prevent the growth of mold.

● After conditioning, pasteurize those foods that require it in the oven to prevent decay from bacteria, insects, or moisture.

● Pack dried food in sealed, airtight containers and store in a cool, dark, dry place.

● If moisture is discovered after the food has been stored, pasteurize the food again, pack in a new container, reseal, and store.

Equipment for Drying

Most of the equipment for preparing and pretreating produce should be readily available around the house. The knife used to cut and prepare foods should be stainless steel to prevent food discoloration. A large kettle or steamer with a wire basket, colander, or open mesh bag is used to blanch or steam vegetables.

Other equipment for drying and packing that may have to be purchased or built includes:

1) Trays or racks on which to spread foods to dry.

2) A piece of clean, loosely woven cloth or netting that is a bit wider and longer than each tray to cover the drying produce.

3) For oven or controlled heat drying, a thermometer that measures as low as 100°F.

4) An oven can be used for controlled heat drying, or one of the home-built dryers described below can be constructed. The average home oven will accommodate approximately 6 pounds of produce for drying at one time.

5) Containers such as jars, cans, and plastic bags and bottles in which to store dried products.

The equipment needed for sulfuring is described on pages 22–25.

With advance planning, trays can be built for use in either controlled heat or sun drying, as well as for sulfuring. Tray dimensions should be determined by the size of your oven. The trays should equal the width of the oven less a slight amount for clearance, and the length of the oven less three or four inches to allow staggering that permits hot air to rise in a zig-zag pattern. Trays used in sun drying should also fit in the oven in case the weather changes and drying must be finished indoors.

Trays should be a convenient, easy-to-handle size—about one to two inches deep, twelve to sixteen inches wide and sixteen to twenty inches long. Trays should all be the same size to allow for stacking during drying and sulfuring, as well as for convenient storage.

Make trays completely out of wood. The tray bottom must be open enough to allow air to pass through for rapid, efficient drying. *Wire mesh should not be used on tray bottoms because galvanized screen is treated with zinc and cadmium, both of which are dangerous when brought into contact with acidic foods.* Aluminum screening discolors and corrodes easily, copper destroys vitamin C, and vinyl-coated screening may not withstand the heat of oven and other controlled heat dryers.

Tray bottoms are best made of thin wooden slats or doweling placed 1/4 to 1/2-inch apart and covered with cheesecloth, mosquito netting or any other cloth netting with a mesh smaller than 1/2-inch to prevent foods from slipping through or sticking.

Lightly coat slats or dowels with an unflavored vegetable oil to keep the bottom clean and to facilitate cleaning after drying. After each use clean the bottoms with hot, sudsy water and a stiff brush, rinse well in clear water, and dry thoroughly to prevent mold. Store trays in a way that will provide good ventilation and keep them clean.

Following are diagrams and instructions for building solar, oven, stove-top, and electric dryers.

Solar Dryer—A solar dryer can be easily constructed from a 3 x 6-foot storm window, although a custom size can be purchased or made if home-built trays cannot be accommodated under the standard 3 x 6-foot glass. The lumber used for the legs, braces, and glass supports is 1 x 4-inch and the tray supports are made of 1 x 1-inch wood. Cut the legs to the proper length, nail the tray and glass supports to them, then nail a brace to the opposite side of the legs. The front pieces can be fastened with screws if the dryer is to be dismantled for storage. Each front piece should be positioned to prevent the window and trays from sliding out. The distance between the window and the tray supports should be determined by the height of the drying trays plus three to four inches to allow air to circulate freely. The sloping side of the the box is placed toward the sun for the best effect.

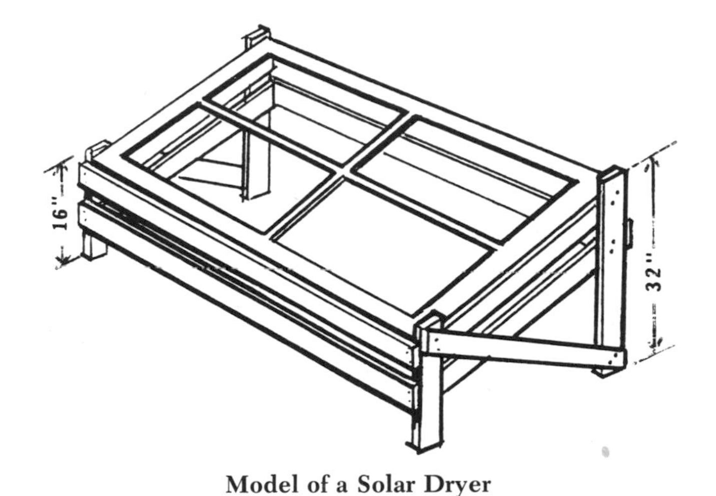

Model of a Solar Dryer

Oven Dryer—An oven dryer facilitates controlled heat drying, but is not an absolute necessity. The principal reason for building one is to increase the drying capacity in an oven with few racks and provide a lever to hold the oven door open to maintain proper temperatures.

Since all ovens are not the same size, modify the dimensions in the model illustration to suit your oven. The lumber for the legs and cross supports is 1 x 2-inch, for the tray slides use 1-inch square lumber; for the door holder use 1/4-inch or 1/2-inch plywood cut with slanted notches the width of the door and spaced about an inch apart. The dryer should be placed on an oven rack in the lowest position and supported by the rack or a small cookie sheet.

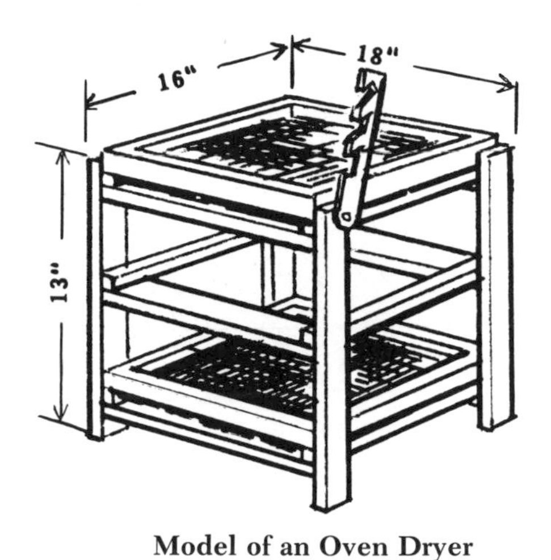

Model of an Oven Dryer

Stove-Top Dryer—A stove-top dryer frees the oven and/or adds additional drying capacity. It can be used over any burner, gas or electric.

The sides and door are framed with 1 x 4-inch wood to which 1/2-inch insulated building board or homosote is attached. The top and back are also insulated building board nailed to the outside of the frame. The inside tray supports are 1-inch square wood. The sheet metal on the bottom extends beneath the frame by two inches and goes completely around the box. The heat dispersing baffle, an 18 x 24-inch piece of sheet metal, is hung horizontally below the lowest shelf with wire attached to screw eyes. The trays must clear the sides of the dryer and be four inches shorter than the dryer itself to allow for staggering when the dryer is being used.

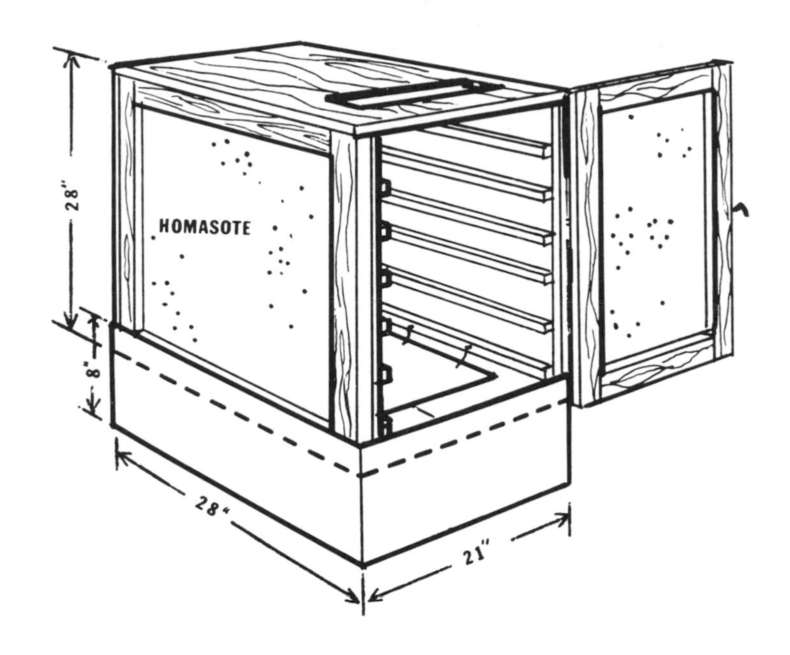

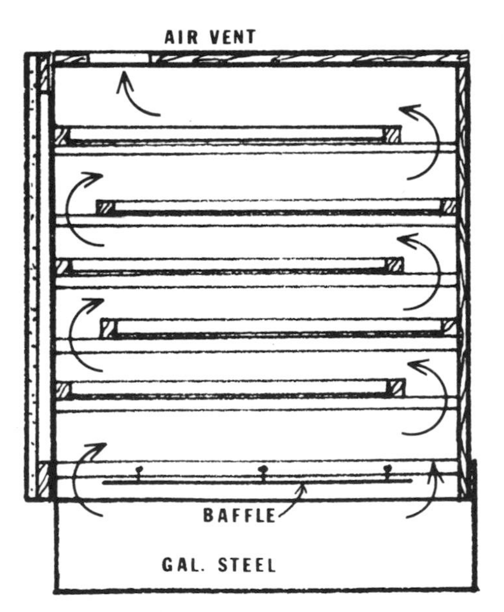

Model of a Stove-Top Dryer

Electric Dryer—An electric dryer is similar to the stove-top dryer, except that it uses light bulbs for heat, a fan for ventilation, and a thermostat for temperature control. The porcelain light bulb sockets, thermostat, and fan are mounted on 1/2-inch plywood supported by two pieces of 1-inch square wood. The fan should have a six-inch diameter blade and a low-speed motor; the bulbs are 200-watt and the wiring should be under the board.

The door is hinged on one side and has a vent cut into it to allow air to circulate. The vent is covered with a slide made of 1/4-inch or 1/2-inch plywood that can be moved back and forth.

The trays should be 4 inches shorter than the box and should be staggered. No tray should be placed on the sheet metal heat disperser.

Produce Selection and Preparation

Select good quality produce for drying. Poor quality fresh food yields poor quality dried food. A rule-of-thumb to apply when selecting food for drying: if you wouldn't serve the fresh food to an honored guest, don't dry it.

Fruits and vegetables should be at peak maturity, free of bruises and decay, fresh, ripe, firm, and perfectly clean.

Wilted or overheated vegetables will not make good dried vegetables. Immature fruits and vegetables will have weak colors and flavors. Overly mature vegetables are often tough and fibrous. Overly mature or bruised fruit is likely to spoil before drying is completed. Fruits or vegetables with just one spot of decay can impart a disagreeable flavor to an entire batch of dried produce.

Home-harvested food is ideal for drying since it is fresher than store-bought and will yield a better quality dried food. Gather the food in the cool, early morning, begin drying immediately, and proceed until finished.

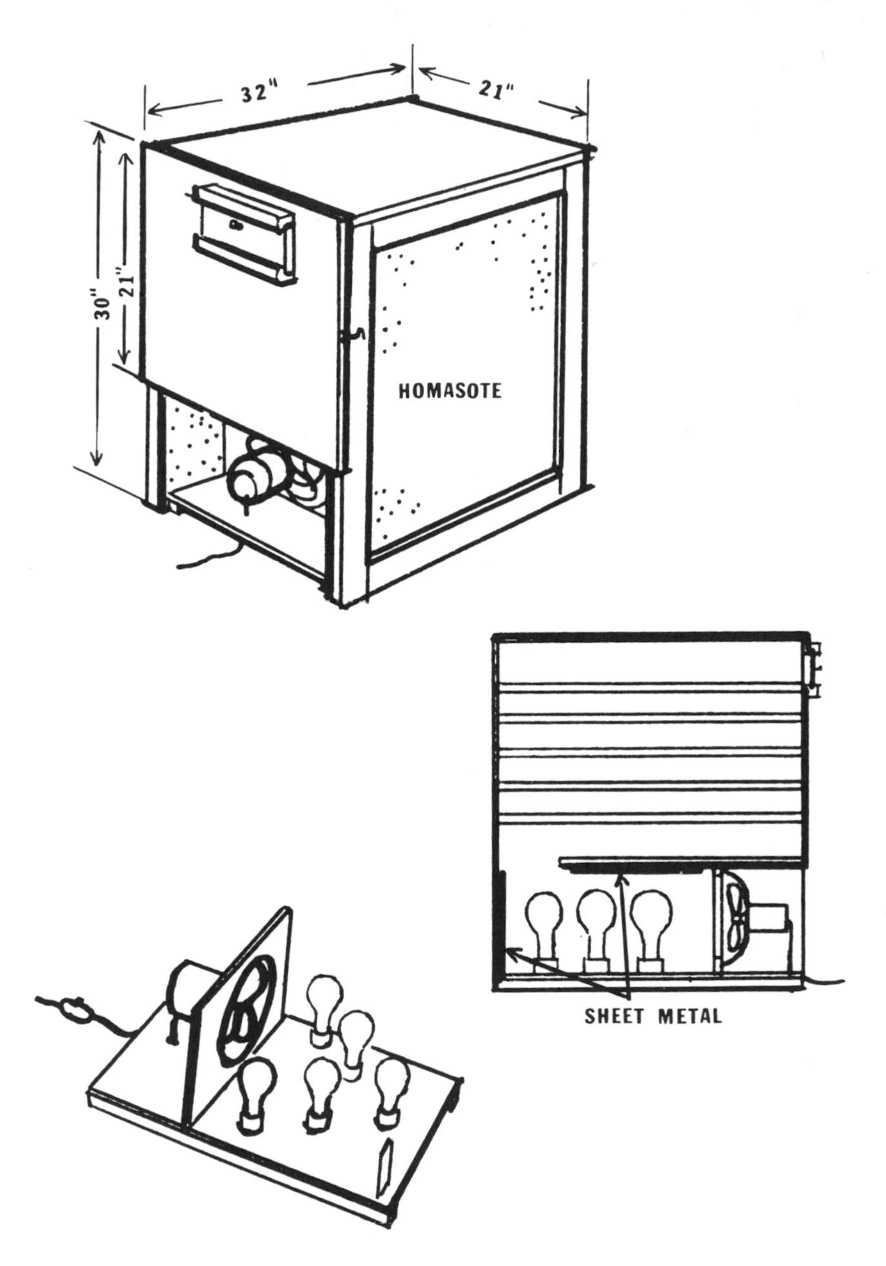

Model of an Electric Dryer

A few fruits need only to be washed and placed on drying trays, but these are the exceptions. All vegetables and most fruits must be pretreated. The goal of pretreatment is to halt enzyme action in vegetables and fruits, and to prevent discoloration and retard other kinds of spoilage in fruits. Pretreatment varies for fruits and vegetables. Each is discussed separately below.

Before pretreating food, clean it if necessary and prepare it according to the instructions for the individual food in the alphabetical directory at the end of this section. Generally speaking, fruits and vegetables should be peeled, cored, sliced, cut, shredded, halved, or quartered before pretreating. But follow the specific instructions for each in the directory.

Do not prepare more produce than can be dried at one time because it will deteriorate rapidly if stored between pretreatment and drying.

Pre-Drying Treatment for Vegetables

Steaming is the preferred pretreatment for vegetables to be dried. Steaming ends enzyme action, helps retain the vegetable's water-soluble vitamin and mineral content, sets the color, speeds drying by relaxing the tissues to make it easier for moisture to escape, prevents undesirable flavor changes during storage, and insures satisfactory rehydration during cooking while lessening the amount of soaking time needed before cooking.

Nutritionists recommend steaming rather than parboiling because steamed vegetables are suspended over boiling water, preserving water-soluble vitamins and minerals. Vegetables immersed in boiling water lose some of these nutrients. As a result, dried vegetables that have been pretreated by parboiling are nutritionally inferior to those pretreated by steaming.

To steam vegetables, suspend them over boiling water

for the specified period of time. For successful steaming, the vegetables must not come into contact with the boiling water and must be loosely packed so the steam can circulate and make contact with all of the pieces.

Use a kettle or other utensil with a tight-fitting lid. Steam or pressure cookers with the petcock open are ideal. Vegetables can be placed in a wire basket, colander, or sieve and supported by a rack. They can also be suspended in a piece of clean cheesecloth or any other loosely woven white cotton material and supported by inverted custard or coffee cups.

Add at least 1/2 inch of water to the bottom of the utensil, put the rack or inverted cups in place, and bring the water to a brisk boil. Then place the vegetables in the pot and cover.

The amount of steaming time varies because vegetables differ in size, shape, ability to conduct heat, and natural enzyme levels. Instructions for each vegetable in the directory specify steaming time.

Steam the vegetables until each piece is heated through and wilted. Test to see if steaming is finished by taking a piece from the center of the batch and pressing it. It should feel tender, but not completely cooked.

Remove the vegetables from the steamer, absorb surface moisture on towels, place on trays, and begin drying.

Pre-Drying Treatment for Fruit

Most fruits begin to discolor as soon as they are cut or peeled. The enzyme action in fruits, while slower and less intense than in vegetables, must also be halted. Pretreatment of fruits, unlike that of vegetables, serves more to improve the taste and appearance of the finished product than combat spoilage.

The most effective and widely used pretreatment method for fruit is also the most widely debated—sulfuring. Apples, apricots, nectarines, peaches, and pears discolor more rapidly than other fruits and are the fruits most regularly sulfured to prevent discoloration and help preserve the dried product. Commercially dried fruit sold in supermarkets is usually sulfured to improve appearance, taste, and shelf-life. Unsulfured fruits may be purchased in health food stores.

The arguments for sulfuring conclude that it decreases the loss of vitamins A and C while preserving color, flavor, and reducing spoilage caused by microorganisms and insect infestation. All sun dried fruits must be sulfured since they are exposed to the air for a much longer time, which results in increased discoloration and insect activity. Sulfuring advocates claim sulfuring does not damage health because the heat of drying and the subsequent cooking dispels the sulfur used in the process.

Sulfuring opponents argue that sulfur destroys any thiamine in the fruit and that the amounts of vitamins A and C preserved by sulfuring are somewhat insignificant when compared with other sources of these vitamins. While sulfur is a mineral needed to sustain life, sulfuring opponents argue that since there are natural ways to obtain it the chemical should not be added to foods.

The decision to sulfur is an individual one. If sulfuring is not done, however, the fruit must be carefully protected from insects throughout the drying process, as well as conditioned and pasteurized afterwards to assure its quality. Unsulfured dried fruits must be stored carefully and checked often for spoilage and insect infestation.

Fruit must be temporarily treated as soon as its flesh is cut to prevent the oxidation that causes discoloration. This applies to fruit that will be sulfured as well as to fruit that will be dried unsulfured. The simplest temporary treatment is to immerse cut fruit for 10 minutes in a solution of four to six tablespoons of salt dissolved in a gallon of water. Then drain, but do not rinse, and dry any surface moisture. This method has its disadvantages, however. First of all, water is added to a fruit from which moisture will later be extracted, thus extending the drying time. Secondly, many of the vitamins and minerals in fruits are water-soluble and are leached out by soaking.

A better method is to coat each piece of fruit with a solution of ascorbic acid, which is vitamin C in crystal form. Pure ascorbic acid is available in drug or health food stores. The specific amount of ascorbic acid and water recommended for different fruits varies. For apples, dissolve 2 1/2 teaspoons of crystalline ascorbic acid in each cup of cold water. For peaches, apricots, pears and nectarines, dissolve one teaspoon of pure crystalline ascorbic acid in each cup of cold water. One cup of this solution will treat approximately five quarts of cut fruit. The solution should be sprinkled over the fruit as it is peeled, cut, diced, sliced, pitted, or cored, making sure that every part of the fruit is adequately covered.

Lemon juice, high in ascorbic acid, may be used as a substitute in an emergency. Commercially available anti-oxidant mixtures containing vitamin C may also be used, but they are not as effective as pure ascorbic acid. Follow the directions carefully for the particular product.

If the fruit will not be sulfured, a saline dip or ascorbic acid treatment is all that is necessary before drying. You may, however, prefer to use one of the alternatives described after the directions for sulfuring.

Cracking Fruit Skins

Some fruits that are to be dried with their skins—peaches, plums, apricots, etc.—require cracking the skin to allow moisture to escape. Other fruits—grapes, prunes, small dark plums, cherries, figs, and firm berries such as blueberries—have relatively tough skins with a waxy coating. These fruits should be dipped in briskly boiling water to remove the waxy substance and crack the skins. They should be held in the water for 30 to 60 seconds, depending on the toughness of the skin, the maturity of the fruit, and the altitude. Afterwards, wash the fruit in ice cold water to chill it quickly. Then, drain and pat dry with a towel.

Some older methods recommend the use of lye (caustic soda), but more recent knowledge rules this out because of the danger involved in handling lye and the damaging effect it has on B vitamins and vitamin C.

To peel the skins—really only necessary for apples—dip the fruit in boiling water for 30 to 60 seconds as described above, chill in cold water and the skins will easily slip off by hand.

Sulfuring

In addition to trays, the following equipment is required for sulfuring: an airtight cardboard or wooden box large enough to fit over the stacked trays of fruit to be sulfured, the sulfur itself, a container to hold the sulfur, and matches.

The sulfuring box should be prepared by cutting a slot three to six inches wide and one inch high in the bottom of the inverted carton to provide air ventilation to keep the sulfur burning. If the carton is cardboard, fold the slot flap up, rather than cutting it off completely so that it can be folded down once the sulfur has burned. Make a small slash or opening near the top of the inverted box on the side opposite the bottom slot. This also helps the sulfur burn, but should be closed once the sulfur is consumed.

Stack wooden drying trays with 1 1/2 to 3 inches of air space between them, using small wooden blocks to keep the trays apart so the sulfur can circulate freely. Elevate the bottom of the stacked trays at least 4 inches off the ground.

The sulfuring box should be high enough to fit over the stacked, elevated trays and leave additional space at the top—preferably 6 inches and not less than 2 inches. The box should be large enough to fit over the trays and to leave room on one side for the sulfur container plus at least another 1 to 1 1/2 inches for clearance and air circulation.

Once the fruit is cut and treated with ascorbic acid or saline solution place it on the trays. The trays can be those used in drying as long as they do not have any galvanized screening, metal, or nails. Sulfur reacts with the galvanized parts, imparting an unpleasant taste to the food and endangering its safety.

Use "sublimed sulfur" or "flowers of sulfur," both available at drug stores. *Do not use garden dusting sulfur.* The sulfur should be marked 99 1/2 percent pure. It is a light yellow powder with no taste but a very slight scent. The scent is nothing at all like the rotten egg odor given off by hydrogen sulfide. When the sulfur burns and becomes sulfur dioxide, the fumes will be irritating to the eyes and nose—so much so that too much inhalation, which could be dangerous, is unlikely.

Even so, it is always best to sulfur outdoors to prevent any problems with fumes. Before sulfuring, weigh the fruit to be sulfured to determine the amount of sulfur needed. The basic ratio is one or two teaspoons of sulfur per pound of fruit if the sulfuring time is less than three hours, and three teaspoons per pound of fruit if the sulfuring time is three hours or longer. Instructions in the directory for individual fruits list the sulfuring time for each.

Spread the fruit in a single layer on the trays. Fruits being dried with the skins on should be placed skin side down to prevent any loss of the fruit's juices. Don't sulfur more than one type of fruit or one uniform size of fruit at any one time; you cannot move the fruit around once the fumes are dispersed in the box to compensate for the different sulfuring times required by different fruits or different sizes of the same fruit.

Stack the trays. Place a container to hold the sulfur to one side, not directly under the trays, and fill the sulfur container to a depth of approximately one-half inch. The container can be made of metal. Although metal will corrode after a number of uses, it is fine for several sulfurings and can then be replaced. The container should be fairly shallow, not much more than 1 1/2 inches higher than the top of the sulfur to enable the chemical to burn freely. If the sides are higher, the sulfur will smother from lack of oxygen. In addition to metal, enameled containers and heavy crockery can be used, but bright red or yellow ceramic containers should be avoided because they may contain cadmium which reacts with the sulfur to produce harmful fumes.

Place the box over the trays and the sulfur container, open the lower flap and light the sulfur with a match. The sulfur will melt first, then burn with a clear blue flame. Do not leave any matches on the sulfur or they may smother part of the sulfur. Do not use paper to ignite the sulfur; unburned pieces may also tend to smother the sulfur and prevent it from burning completely. After about 5 minutes, check to be sure that the sulfur has ignited and is burning.

When all of the sulfur has burned, close the lower flap and seal the opening on the opposite upper side of the box. Sulfur spread approximately 1/2-inch thick in the container will take about 15 minutes to burn, depending on the ventilation, shape of the sulfur container, weather conditions, and other factors. Once the sulfur is completely burned and the box is sealed, start counting the time the fruit is to remain in the sulfuring box. Wait until this point to begin calculating the time so that the sulfur can penetrate all parts of the box and work on the fruit surfaces. Try to prevent any sulfur fume leaks from around the base of the box. Leave the box flaps out and weight them down with rocks, cement blocks, bricks, or earth to provide an airtight seal.

Fruit tends to take on a uniform translucent appearance and the juice begins to ooze out when the sulfuring is finished. Smaller pieces and more mature fruit require less sulfuring time.

When the prescribed sulfuring time has elapsed, face the box with your back to the wind so the fumes won't hit you in the face as you slowly tilt the box toward you and lift it off the stacked trays. Let the fumes escape downwind.

If the fruit is on trays that will be used for drying, place the trays in the sun or the controlled heat dryer. If the fruit must be transferred to other trays, be careful not to spill any of the juice that may have collected in the fruit.

Alternatives to Sulfuring

Sulfur treatment may also be accomplished by immersing the cut fruit in a solution made with sulfur compounds and water. The instructions for this process vary. The chemicals used include sodium sulfite, sodium bisulfite, or potassium metabisulfite. The strengths vary between one teaspoon of chemical per pound of fruit in a gallon of water to three ounces of chemical in five gallons of water. Soaking time varies according to the fruit—10 to 15 minutes for apricot halves, 15 to 20 minutes for peach halves that have been steamed and cooled to room temperature and then soaked, 20 to 25 minutes for pear halves. *We do not recommend soaking in a sulfur solution* because many of the water-soluble vitamins, minerals, and sugars will be leached out in the process. In addition, the fruit absorbs water which extends the overall drying time.

Saline solutions and ascorbic acid solutions, described as temporary treatments before sulfuring, are considered alternatives to sulfuring as well.

Fruits may also be steamed instead of sulfured. They are steamed in the same manner as vegetables. The appearance and flavor of steamed fruits that are later dried is different from and, some believe, inferior to sulfured fruit. Steaming is not recommended for fruits that will be sun dried. Sun drying exposes fruits to open air and, since these fruits have taken on added moisture, they will be exposed for a longer time inviting problems with insect infestations. Dried steamed fruits generally have a darker color than dried sulfured fruits. Apricots, peaches, and pears may pick up a slightly cooked taste if steamed. Steamed fruits may also be soft and difficult to handle.

The final alternative to sulfuring is syrup blanching. Dried fruits are nearly 75 percent sugar by weight. Syrup blanching adds more sugar and creates a very sweet, almost candied, product. But it helps significantly to maintain an appetizing color in unsulfured apples, apricots, figs, nectarines, peaches, pears, and plums.

Make a syrup using one cup of corn syrup and one cup of sugar to three cups of water. Or, depending upon the sweetness desired, use any of these combinations: one part corn syrup to one part water; one part sugar to one part water; one part sugar to three parts water. Syrups with more water and less sugar or corn syrup will produce treated fruits that are less sweet.

Heat the syrup to boiling, 212°F, add the fruit and simmer for 10 to 15 minutes, according to the size of the fruit pieces. Remove the container from the stove and let the fruit stand in the syrup for another 10 or 15 minutes, again depending on the size of the fruit pieces. Remove the fruit from the syrup, drain it on paper towels to remove as much moisture as possible, and cool.

Fruit treated this way will take longer to dry because the syrup adds moisture. The sweet syrup will also attract insects so the fruit should be well protected if dried outdoors and should be stored in a very cool, dry place.

During the drying process, be sure that syrup blanched fruit doesn't stick to the trays or scorch while dried in a controlled heat dryer. Line the tray bottom with cheesecloth.

Methods of Drying Produce

Both sun and controlled heat drying have advantages and disadvantages. Sun drying has no energy cost, controlled heat requires electricity or gas. Most agricultural extension services and food technologists have found that controlled heat dried foods are superior to sun-dried foods in color, flavor, cooking quality, and nutritional value. Controlled heat dried foods are also more likely to be free from insects and insect eggs than those dried outdoors. Controlled heat drying takes less time than sun drying and can be done without regard to the weather.

Successful drying depends on achieving three goals: 1) halting chemical changes with proper pretreatment as soon as food is cut into pieces; 2) using a temperature high enough to prevent microorganism growth that spoils food, but not a temperature so high that food cells burst, spill their juices, or cook or scorch the food; 3) having good circulation of dry air all around the produce as it dries. These conditions can be met outdoors under the sun or created artificially indoors in a controlled heat dryer.

Sun Drying

The ideal climate for successful sun drying provides a brightly shining sun, dry air, and a good breeze to carry the moisture away. All of these conditions must be present at the time the food to be dried matures. Some parts of the United States have better climates for sun drying than others.

The best areas for sun drying are the southwestern states and the inland parts of California. The Plains States rank second. The humidity in southern states works against the sun's heat, although some successful sun drying has been done there. Sun drying has been done successfully in other parts of the country when climate conditions are ideal. Sun drying should not be attempted in smog areas or anywhere near a highway because the smog or engine exhaust fumes tend to impart an undesirable flavor to the dried produce.

After pretreating the food that is to be sun dried, place it in a single layer on drying trays in a way that will permit air to flow freely above and below it. Cover the trays with cheesecloth or any other loosely woven cotton cloth to protect the food against insects, rain, dew, and dust. The covering should not touch the food.

Place the trays in the shade for at least an hour before

putting them in direct sunlight. Very hot, direct sunlight may dry fruit and vegetable skins too rapidly making them tough. In such instances, the drying should best be done in the shade. Trays may be positioned flat or tilted toward the sun, but not set directly on the ground. A gently sloping roof facing south is a good location for direct sunlight drying. Trays can also be supported on tables, saw horses, stones, or wooden blocks. The important factor is that the trays be far enough above ground so that air circulates freely over and under the food and possible damage from dust and insects is reduced.

Produce should not be placed outside for drying until the morning dew has evaporated and should be taken in at night before the evening dew begins to form. If the night air will be very dry, and if the temperature is not going to drop more than 20°F below the day's high, produce may be left outdoors overnight. Otherwise, it should always be taken inside at night.

Several times during the day, turn the food gently to help it dry evenly. Wait until any juices have dried before turning sulfured fruit. Take the food outside for two or more succeeding days if necessary. As the food is nearly dried, there is a danger it will be scorched by a hot sun. Combat scorching by turning the food more frequently to reduce the temperature, moving the food into the shade, or by stack-drying.

To stack-dry produce, place it in direct sunlight for a day or so until it is two-thirds dry. (To determine this, weigh the produce after pretreating. When the same batch after drying weighs about half of its beginning weight, the produce is two-thirds dry.) Then stack the trays one on top of another in a place away from direct sunlight, but in the path of very dry, strong breezes. This method will not work in damp, muggy, humid climates. Keep about six inches between the trays and continue until the produce is dry.

The amount of time required for sun drying depends on the temperature, humidity, and the size of the fruit or vegetable pieces. Drying smaller pieces of produce in a warm, dry climate takes one or two days. Drying larger pieces of produce in cooler, damp climates may take as much as seven to ten days.

The solar dryer described in the equipment section is used in much the same way except that produce in a solar dryer should be checked more frequently, about once every hour or so, to prevent uneven drying and scorching. The solar dryer is more efficient if turned to keep it constantly facing the sun.

Instructions in the directory section for individual fruits and vegetables describe how to test when each is dry. Once the produce tests dry, package, condition, pasteurize and store according to the instructions given for those procedures.

Controlled Heat Drying

Artificial, controllable heat speeds up the drying process by using temperatures just short of those that would cook the food. Controlled heat drying is faster than sun drying and the end product is often better in terms of quality and flavor.

Most indoor drying is done in electric or gas ovens. As much as 20 hours of oven time may be required. If the oven cannot be spared for that much time, build one of the controlled heat dryers described earlier. The following instructions refer to ovens, but the same general principles apply to any controlled heat dryer.

Rapid drying produces the best product. Moisture will evaporate and be carried from the oven faster with more heat and greater air circulation. Food should dry from the inside out. If the temperature is too high, the outside will harden and prevent inside moisture from escaping.

If the temperature rises above 175°F, the cell walls in the produce will break, the juice will leak out, and the produce will cook rather than dry. The best drying temperatures range between 110°F and 160°F.

Generally speaking, the drying temperature should be maintained between 140°F and 150°F for the best results. Start with low temperatures (110°F–130°F), gradually increase to middle and higher temperatures (140°F–160°F) and decrease to low temperatures when the produce is two-thirds dry to prevent scorching or burning. At least half of the time the food is being dried the temperature must be 140°F or higher to kill the microorganisms and remove the moisture that can create mold.

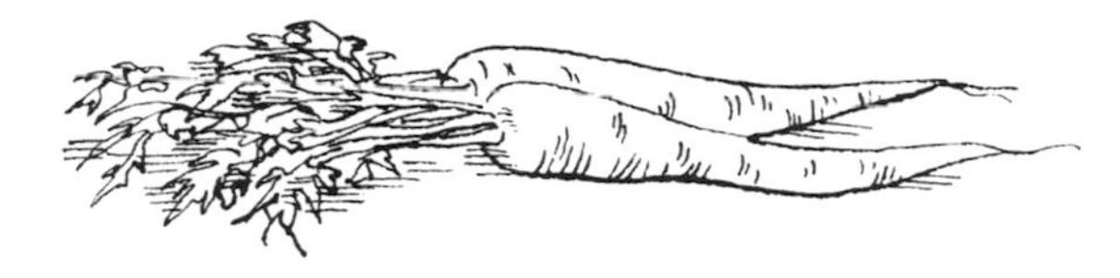

Preheat the oven to its lowest temperature. Turn off or remove the top heating element in electric ovens. If the minimum heat temperature is above what is required, open the oven door to lower it. With electric and newer model gas ovens you will probably have to prop the door open only about 1/2 inch to reach the desired temperature. The doors of older model gas ovens may have to be opened as much as eight inches.

Arrange the produce evenly on drying trays. Lighter loads and smaller pieces will dry faster. Most ovens can dry four to six pounds of fresh, pretreated food at one time. Spread the produce in a single layer; fruits should be skin side down so the juices will not escape.

Arrange trays in the oven allowing a three-inch space at top and bottom. Trays should be 2 to 2 1/2 inches apart. Use an oven dryer, wooden blocks, or doweling to stack trays if you don't have enough oven racks.

Stagger the trays, front and back alternately, to create a zig-zag air circulation. Place oven thermometers on top and bottom racks to check the temperature. Food may be scorched if heat accumulates at the top because the oven door isn't open wide enough, or at the bottom because the oven temperature is set too high.

After drying begins, check the food every half hour. Rotate trays from top to bottom and turn them from front to back to speed even drying. Pieces on the sides will dry faster than those in the middle. To compensate for this difference, shift the food around on the tray every half hour, or remove pieces as they dry. Large pieces should be turned over when half the drying time has elapsed, but fruit halves with juice in them should not be turned over until the juice has evaporated.

Food drying in the oven at the proper temperature and rate should be moist when touched and cooler than the air flowing around it. If the food is as warm as the surrounding air and is not moist, it is drying too fast. Lower the temperature or open the oven door to increase ventilation.

Do not use any type of cloth to protect drying trays in a gas oven. If you use a stove-top dryer, do not boil water, cook, or do anything that will raise the temperature or humidity around it. By all means, do not attempt to dry food over a floor register because the food will be contaminated with dust and may also absorb heater fumes that will spoil it.

Is It Dry?

The time required to dry particular foods varies with the type of food, the size of the pieces, the drying method and equipment used, the temperature, air circulation, and humidity. As a general rule, the less time taken to dry a fruit or vegetable, the higher the nutritional value of the dried product and the better its color, flavor, and texture.

Dried fruit should not have more than 20 percent moisture, and vegetables not more than 10 percent, to be stored without spoilage or deterioration. Not all pieces dry at the same rate. Remove pieces as soon as they are dry.

Drying times and tests for dryness are different for fruits and vegetables.

Fruits — Fruits generally take between 6 to 24 hours to dry in a controlled heat dryer, and as much as 10 days in the sun. To test fruit dried under controlled heat, remove it from the dryer and allow it to cool for at least 10 minutes. Hot fruit always appears softer and moister than cooled fruit. Sun and controlled heat dried fruit should not be sticky, although some fruits like figs and cherries will feel slightly sticky when dry. Press a handful of fruit together. When dry it will spring back to its original shape and the pieces will separate from one another. If the pieces shatter the fruit is too dry. Properly dried fruits are pliable, leathery, suede-like, or springy. Cut open one of the larger pieces of fruit through the thickest section. If it is properly dry, no moisture will ooze out and the inside will have the same texture and appearance as the outside.

Vegetables — Drying times for vegetables vary between 3 and 15 hours in a controlled heat dryer, and may take as long as several weeks to dry in a warm, dry room. Vegetables are rigid, brittle, hard, crisp, and bone dry when they are finished. Some vegetables such as corn, beans, and peas will rattle if shaken on the drying tray. If you are in doubt about whether a vegetable is finished in a controlled heat dryer, lower the temperature and leave it in a bit longer. As long as the temperature is low, extra time in the dryer will not damage the vegetable.

Approximate Yields of Dry Foods from Fresh

Twenty-five pounds of fresh fruits such as apples, peaches, and pears will yield between three and five pounds of dried fruit. Twenty-five pounds of fresh vegetables will yield two to three pounds of dried vegetables. Twelve big ears of fresh corn will yield about one pound of dried corn. The different fruit and vegetable yields are due to different acceptable moisture levels in the dried products. Dried fruits can contain twice as much moisture as dried vegetables.

Conditioning and Pasteurizing

Dried foods cut in different sizes dry at different rates. Several pieces of dried product may appear completely dry while others in the same batch still contain some moisture. Since moisture is the breeding ground for mold it must be reduced. The moisture in dried foods can be equalized by a process called conditioning.

To condition produce, place the cooled food in an open, nonporous container that is not made of aluminum. The container might be crockery, glass, or an enamelized metal. It should be kept in a warm, dry room where air circulates freely and where there are no insects or animals.

Stir the food once or twice a day. Conditioning will take anywhere from 3 days to 2 weeks, depending on the size of the food pieces. Corn is an exception—it can be conditioned in a day or two.

After conditioning, foods should be pasteurized to kill any insects or microorganisms they may have picked up while exposed to the air. Pasteurization can best be done in the oven. Spread the food on trays to a depth of not more than an inch. Preheat the oven to a temperature of 150°F and place the trays inside for 30 minutes. (In a

175°F oven food will pasteurize more rapidly; fruits require only 15 minutes at that temperature and vegetables only 10 minutes. *But be careful that the temperature does not rise above 175°F or the food will cook rather than pasteurize.*)

Package the food immediately after pasteurizing to prevent any bacteria from contaminating it and any moisture from being re-absorbed.

Packing and Storing

Opinions differ as to whether food should be cooled before packaging after it is dried. Packaged warm, the food may tend to sweat; allowed to cool it may pick up bacteria and moisture. There is no disagreement, however, that if dried food is cooled before packaging it should be protected from bacteria, moisture, and insects while cooling.

Dried food can be stored in any container that is moisture-proof, insect-proof, and bacteria-proof—in other words airtight. Glass jars, coffee or other tin cans with tight-fitting lids, plastic bags, and moisture-proof freezer cartons all make good storage containers. Don't store sulfured fruit in any metal container without first placing the fruit in a plastic bag. Transparent containers should be wrapped in brown paper to prevent light from entering, since light can shorten the storage life of dried foods. Glass containers have an advantage over the others because condensation forms inside to warn you when the moisture level has risen.

Package dried foods in portions that will be used at one time. Opening and closing a container allows moisture to enter which can be absorbed by the food and cause mold. Seal containers so they are airtight. A one-inch strip of clean cloth dipped in hot paraffin and wrapped while warm around the outside of the lid where it joins the container will create an airtight seal.

Packaged dried foods should be stored in a cool, dark, dry place. Warmth, light, and moisture will spoil dried foods.

Examine the stored dried foods often for spoilage or moisture. Discard all spoiled foods. When moisture appears, rescue dried food by repasteurizing and repacking in a new container. Examine dried foods for moisture after an especially long rainy season.

Because of their high sugar content, dried fruits will keep longer than vegetables. Most dried fruits can be stored for about one year. Vegetables, especially green ones, lose their texture and flavor more rapidly. They should be used within 6 months at the most.

Reconstituting and Cooking Dried Foods

Dried fruits such as pineapples, peaches, apricots, figs, dates, apples, and pears can be eaten without rehydrating or cooking. Properly dried foods will return *almost* to their original size and shape when reconstituted and cooked, and retain much of their original aroma, flavor, vitamins, and minerals.

Reconstitution or rehydration consists of returning to the product approximately the same amount of moisture removed during drying. The process differs for fruits and vegetables.

Most fruits should be soaked in water for one to eight hours—the length of time depending on the type of fruit, the size of the pieces, and whether hot or cold water is used. Rehydration takes longer in cold water than in hot water.

Immerse the fruit in just enough water to cover; more can be added later if needed. Too much water will leach out vitamins and minerals from the dried fruits. Don't oversoak the fruit or it will lose flavor and may become mushy and water-logged. Don't soak the fruit too long or it may ferment. Cover rehydrating fruit with a tight-fitting lid.

To test rehydrated fruit, taste one piece to see if it is tender. Longer soaking may result in more tender fruits; soaking too long may result in mushy fruit.

To cook the fruit, simmer it in the same water used for rehydration to help retain the nutritional value. If the fruit is to be used in a pie or dessert, put it in a container, barely cover with boiling water, and soak for several hours. Use the soaking water as natural juices.

Less sugar is required when cooking with dried fruit because some of the starch in the fruit turns to sugar during the drying process. Don't add sugar until after the fruit is cooked. Sugar added earlier interferes with the absorption of water and tends to toughen the fruit fibers.

A pinch of salt helps bring out the flavor and natural sweetness of most fruit. Add a few drops of lemon, orange, or grapefruit juice just before serving fruit to help give it a fresh flavor as well as an extra bit of vitamin C. Cooked dried fruits can be cooled overnight in the refrigerator and served in the same way as canned fruits.

To reconstitute all dried vegetables except greens, barely cover them with cold water and allow them to soak for twenty minutes to two hours or until tender. Greens require no rehydration; simply plunge them into boiling water and cook.

To cook dried vegetables, bring them to a boil in the water in which they were soaked and simmer until done. The amount of water used in soaking and cooking should be the same quantity the food will absorb. One cup of dried vegetables will absorb about two cups of water. It is better to add water later than to start with too much. Any extra water should be evaporated during cooking. Always cook dried vegetables in a pan with a tight-fitting lid to help retain the vitamins and minerals.

Dried vegetables will lose flavor and texture if overcooked. Since dried vegetables have already been partially cooked when they are steamed, they require less cooking time than fresh vegetables.

Some Special Cases

Drying fruits and vegetables is somewhat different from drying herbs for seasoning or tea, preserving flowers, making fruit leathers, and candying fruit. These are special cases, but each can and should be considered under the general category of drying.

Herbs—An outdoor summer garden will provide herbs throughout the year if you grow enough for fresh use in summer and dry the extra harvest for later use during the winter. Drying herbs is much easier than drying fruits or vegetables because herbs do not require extensive preparation or pretreatment.

Herbs may be dried in small bunches suspended upside down in any dry room or area, or in the open over a screen or tray. Both methods are effective and provide a high quality finished product. Herbs can be dried anytime between June and September. They should be harvested for drying just as the buds emerge and before they flower—the time when the plants contain the largest amount of essential oils that give herbs their special flavor and aroma. An especially good time to harvest is the day after an ample rain. Rain rinses herb leaves naturally. If you can't harvest then, rinse dusty or muddy leaves gently in cold water and drain.

Cut the herbs on a sunny morning, just before the sun is high, and after the dew has evaporated. Cut the plants down by about one-third; they will grow back to provide a second harvest. Rinse the cuttings if necessary, drain, and pat dry with towels. Discard imperfect leaves.

Suspend the herbs upside down in a tied bunch, or dry them over a tray or screen. If the herbs are suspended, place the bunch inside a paper bag, taking care that the bag does not come in contact with the herb leaves. Once the herbs are dry, use the bag to crumble the leaves and remove them from the stems. Drying by either method will take a few days. Leaves will crumble easily when drying is finished.

To dry herbs on a tray, pinch the leaves from the herb's stalk, arrange them over a small mesh screen, and dry in a warm, airy, dry room. Place a thickness of cheesecloth between the herbs and the screen or tray to facilitate lifting the dried herbs without breaking them.

Keep herbs being dried by either method out of direct sunlight during and after drying to prevent loss of color, flavor, and aroma.

Pack dried herbs in glass or plastic containers and seal airtight. Do not use cardboard containers or paper bags because they absorb the essential oils you want to preserve. Store in a dry, dark, cool place away from the kitchen stove.

Flowers — Dry flowers for use in teas by the screen method described above for herbs. For a sweeter product to accompany desserts or for after dinner, candy or sugar flowers.

To candy flowers, break an egg white into a small bowl and stir but do not beat it. Stir in a teaspoon or two of cold water causing as little foam as possible. With a small paint brush, coat each side of every petal and leaf with the mixture and immediately dip or dust the flower with granulated sugar, coating it thinly but evenly.

Place the sugared flowers on a rack covered with aluminum foil. Turn the flowers after one day and let them stand for another day or until completely dry. Other methods for treating specific flowers are given in the directory that follows.

Teas—The leaves, flowers, roots and bark of a wide variety of plants, herbs and trees may be dried and used to prepare tea. The teas prepared from them are not only tasty but also healthful and soothing. Drying may be readily done by the screen method described above. To brew, allow one heaping tablespoon of the dried tea mixture to steep for about 15 minutes in a pint of water brought to the boiling point. Strain the tea on serving.

Some of the more common herbs and plants used for teas are described in the directory section that follows. There are, of course, many others and the list given here is only a partial one. Among the flowers used to brew tea are: borage, camomile, clover, gorse, hollyhock buds, lime blossoms, marshmallow, orange blossoms, plum blossoms. Plants whose leaves are used for tea include: burnet, costmary, lavender, lemon verbena, pennyroyal, peppermint, rose geranium, spearmint, wintergreen. Both the leaves and flowers of the following plants can be used to make tea: agrimony, betony, catnip, hyssop, tansy, woodruff, yarrow. Roots used for tea include: angelica, burdock, ginseng, licorice, marshmallow, sarsaparilla, sassafras. Among the barks most often used are: bayberry, birch, sassafras, slippery elm.

Fruit Leathers—Fruit leathers, once a specialty item, are now appearing regularly in supermarkets. The leather is pureed fruit with sugar added that has been spread in a thin layer and dried. Fruit leathers are a good way to make use of imperfect fruit rejected in drying, canning, freezing, or making preserves.

Before preparing the fruit, tape strips of plastic food wrap on a cookie sheet or board. Cover fruit leathers to be dried outdoors with a layer of cheesecloth stretched over boards at either end of a table so that the cloth doesn't come in contact with the fruit, but prevents insects from entering.

Wash and prepare the ripe fruit according to the directions given below for each fruit. Cut off any bruised or blemished fruit, then measure the quantity. Use up to five pints for any single batch. Add sugar and heat each fruit as directed. Remove the fruit from the heat and put it in a blender or through a food mill or strainer to puree it. Cool the mixture until it is lukewarm. Then pour the puree on the plastic covered surface and spread to a thickness of 1/8 to 1/4-inch. Five pints will completely cover a 12 × 30-inch strip.

Apricots: Cut in half and remove the pits. Use 1 1/2 tablespoons of sugar for each cup of fruit or 1 cup for 5 pints. Heat to about 180°F or just below the boiling point and crush.

Berries: Remove stems and leave berries whole. Use 1 tablespoon of sugar for each cup of strawberries, or 1/2 cup for 5 pints; 1 1/2 tablespoons of sugar for each cup of raspberries, or 1 cup for 5 pints; and 2 1/2 tablespoons of sugar for each cup of blackberries, or 1 1/2 cups of sugar for 5 pints. Strawberries should be brought to a full rolling boil and removed immediately from the heat and pureed. Other berries should be boiled until the liquid becomes syrupy. Then put them through a food mill or strainer to remove most of the seeds. Berries should be spread to a thickness of 3/16-inch.

Peaches and Nectarines: The best varieties to use are yellow freestones such as Elberta, Rio Oso, and Redhaven. Peel and slice the peaches. Do not peel nectarines. Use 1 1/2 tablespoons of sugar for each cup of fruit or 1 cup of sugar for 5 pints. Crush while heating to 180°F or just below boiling. If the liquid appears thin, boil until it becomes syrupy.

Plums: Use varieties with firm flesh such as Santa Rosa, Mariposa, or Nubiana. Slice the plums and use 2 1/2 tablespoons of sugar for each cup of Santa Rosa plums or 1 1/2 cups of sugar for 5 pints. For other plums, use 1 1/2 tablespoons of sugar or 1 cup for 5 pints. Cook the same as peaches and nectarines.

Fruit leathers dry outdoors in 20 to 24 hours depending on the sun's heat and the type of fruit. Do not allow the drying leather to remain outside overnight. Leathers are dry when they feel firm. A good test is to peel the fruit leather off the plastic. If it comes up completely without leaving any puree on the plastic, the leather is completely dry. Don't leave the leather in the sun any longer than necessary.

If the weather changes, or the humidity is too high, bring the fruit leather indoors, and dry the leather in an oven heated between 140°F and 150°F. Leave the door slightly ajar.

To store fruit leathers, roll them up on the plastic sheets, cover with more plastic food wrap, and seal tightly. Color and flavor will be well preserved for about one month at room temperature, four months in a refrigerator, and up to one year in a freezer.

Candied and Glaceéd Fruit—Fruits that are tender when ripe but have a firm flesh—peaches, figs, pineapples, etc.—can be preserved by candying. The candying preservative is sugar, which is slowly impregnated in syrup form into the fruit until enough is present to prevent spoilage. Water is drawn out of fruit immersed immediately in heavy syrup. Crisp fruits such as apples and soft and mushy berries will not candy well using the following process. Citrus fruit peels require a different process described later.

Prepare the fruit for candying. Wash and cut into pieces of the size desired. Then prepare the first of five syrups in which the fruit is brought to a boil and then cooked for two minutes. Sugar can be used for the sweetener, but it tends to dry the fruit more than a light corn syrup.

The first syrup is made of 2 cups sweetener to 4 cups water; the second is 3 cups sweetener to 4 cups water; the third is 5 cups sweetener to 4 cups water; the fourth is 7 cups sweetener to 4 cups water; and the final syrup is 10 cups sweetener to 4 cups water.

Candying is a five-day process. Each day bring the prescribed syrup to a boil, add the fruit, bring to a boil again and cook for 2 minutes. Remove the container from the stove and let the fruit remain in the syrup for 24 hours. Weigh the fruit down with a dish or other object. Then drain the syrup, and begin the process again the next day with the next syrup.

After the fruit is cooked for the last time, let it stand in the last syrup for approximately 3 weeks. Remove the fruit from the syrup, dip the fruit quickly in boiling water to remove some of the stickiness, then dry on screening at room temperature or in a 120°F oven. Pack candied fruit in shallow airtight containers, placing a layer of waxed paper between the layers of fruit. Store in a dark, dry, cool place.

Candied fruit can also be glacéed after it is dry. Heat a syrup of 3 parts sugar, 1 part corn syrup and 2 parts water until it reaches softball stage (234°F–238°F on a candy thermometer). Cool the syrup to below 200°F and quickly dip the candied fruit in it, drain, and dry. Package and store glacéed candied fruit in the same way as regular candied fruit.

To prepare citrus peels for candying, cut the peel lengthwise in sections, cover with water, and bring to a boil. Cook for about 15 minutes until soft, drain, and scrape out the white. Then cut the peel in thin sections.

Boil a syrup of 1 cup sugar, 1/4 cup water, and 2 tablespoons of corn syrup (omit the corn syrup if the peel is to be hard; corn syrup makes the peel soft and chewy). Then add 1 cup of the prepared peel and cook slowly until the peel is almost transparent, a temperature of about 230°F on a candy thermometer. Remove the peel from the syrup, cool, roll in granulated sugar and dry. Store in an airtight container in a dark, dry place.

DIRECTORY

An alphabetical listing of fruits, vegetables, nuts and herbs that can be dried in home facilities, along with procedures for treating each

Almonds

In the United States, most almonds are grown in states west of the Rocky Mountains, with California the most prominent. Late in the fall the outer hull of the nut splits open and the shells and kernels begin to dry. When most of the hulls in the center of the tree have split, harvest the whole tree by knocking the nuts down with a long pole. Remove the hulls and place the nuts in the shade for three to five days or until dry. Packed in airtight containers, the nuts can be stored in a freezer for as long as a year and at room temperature for four to six months.

Anise

Hippocrates, the father of modern medicine, recommended that anise be grown in every herb garden—as a cough remedy. Today anise leaves are more popularly used to flavor foods and the seeds to flavor salads, cakes, and cookies.

Collect the leaves for drying anytime before the seeds appear. Dry the leaves on a wire mesh, cheesecloth, or other lined tray in a warm, dry, well-ventilated room. Pack the dried leaves in an airtight bottle and store in a cool, dry, dark place.

Harvest the seeds when their tops are a gray-green color, at which time the seeds will separate with ease. Before this time, the seeds will stick together. Wash the seeds with cold water to remove any dirt, spread them on wire mesh, cheesecloth, or brown paper, and place in a warm, dry, well-ventilated room until dry. Shake the seeds around occasionally to make sure they dry completely. Bottle and store.

Apples

Select late maturing apples of good cooking or dessert quality. Make sure, they are mature, but not soft. Varieties that have been successfully dried include Baldwin, Ben Davis, Gravenstein, Newtown Pippin, Northern Spy, Spitzenburg, Winesap, Rome Beauty, Red Delicious, Golden Delicious, Jonathan, and the Russets.

Wash and peel the apples. The peel must be removed because it will not rehydrate as well as the pulpy part of the fruit. Apples with the peel left on will be tough when rehydrated.

The apples can be cored before or after slicing. Slice apples in any uniform size between 1/8 to 1/2-inch. Thinner slices will dry faster. Apples may also be diced in 1/4-inch cubes, quartered or cut into eights. Again, pieces of uniform size are desirable to facilitate both uniform drying and sulfuring.

To prepare the apples for drying, use one of the pretreatment methods described on pages 19–21, either coating the fruit with an ascorbic acid solution or immersing it in a saline solution.

Apples require additional pretreatment. Either of the following methods may be used. Steam blanch the pieces for five to ten minutes, depending on their size. Remove the excess moisture by patting dry with a towel. This method is preferred for apples that will be dried in controlled heat. Or, sulfur the apples, using two teaspoons of sulfur for every pound of cut fruit, for 30–60 minutes, again depending on the size of the pieces. To reduce sulfuring time, steam blanch and sulfur. Fruit that is pretreated in this way should be sulfured for approximately 30 minutes.

Arrange the fruit on drying trays no more than 1/2-inch deep. Sliced fruit may be allowed to overlap.

For oven drying, begin at a temperature of 130°F, gradually increasing the temperature to 165°F and reducing the temperature to 145°F as the drying is nearly finished to prevent scorching. Drying will take approximately six to eight hours under these conditions. After drying, condition the fruit, pasteurize, package and store.

Prepare and pretreat the apples in the same manner for sun-drying. Dry, then pasteurize and package.

Apples may also be dried in a room. String the slices on a piece of clean string and suspend near the ceiling of the attic or any other warm, dry, well-ventilated room. Pasteurize and package.

When dry, apples are soft, leathery, pliable, suedelike, show no moisture when cut and squeezed, have an elastic, springy feeling when compressed, and when pressed into a ball will separate quickly when the pressure is removed, but will still be soft enough to adhere slightly to the fingers.

For variety, dip peeled cut apples in a thin boiling sugar syrup and dry quickly at 150°F. Done in this manner, apples will be as attractive and light in color as commercially available apples.

If beads of moisture appear on apples dried in any manner and they become sticky, the temperature is too high.

To rehydrate, soak one cup of dried apples in two cups of water for six to eight hours.

Apricots

The best varieties of apricots for drying are Blenheim, Royal, and Tilton, but any variety will do. Select fully ripened fruit before it falls from the tree, but not so soft that it is easily mashed or will lose shape during drying.

It is not necessary to remove the apricot skin, but you may if so desired. Cut the apricot in half and remove the pit. Either leave in halves, or cut into slices. Coat with ascorbic acid solution while preparing the fruit.

Pretreat apricots with one of the following methods. Steam blanch the halves for five to ten minutes and towel dry. Or, sulfur the fruit using one teaspoon of sulfur for each pound of cut fruit. Sulfur halves for one or two hours; slices for one hour. Or use syrup blanching for a more candied product.

Arrange the fruit in a single layer on the drying trays, pit side up for the halves. Be careful not to spill any juices that have accumulated in the halves during sulfuring.

For oven drying, begin with a temperature of 130°F, gradually increasing to 150°F, and reducing to 140°F during the last hour or when the apricots are nearly dry. Do not turn the apricots before the juices in the pit hollow have disappeared.

The average drying time for halves is 14 hours in controlled heat, one or two days in the sun. When dry, apricots will be leathery and pliable, a handful of pieces will separate easily after squeezing together, and the pieces will show no moisture when cut in half. After drying, condition the fruit, pasteurize, package and store.

To rehydrate, soak one cup of dried apricots in two cups of water for six to eight hours.

Asparagus

Select stalks with tender tips. Cut off the tips and pretreat them by steaming for four or five minutes. The average oven temperature for drying asparagus is 150°F. The vegetable should be finished drying after six to eight hours, when the tips will be brittle, crisp, and greenish-black in color. To refresh, soak one cup of dried asparagus in 2 1/4 cups cold water for 90 minutes.

Basil

Basil is known to the French as the herb royale. And according to Pliny the Romans believed that the more curses said over the basil seeds as they were sown, the better the plant would grow.

When the basil flowers begin to open, cut off the upper half of the entire plant. This is the point when the aromatic oil is at its pungent best.

The basil leaves can be preserved in a pickle crock. Alternate layers of leaves and layers of coarse salt. When you are ready to use them, remove a batch and shake the salt free.

Dry the leaves by suspending a bunch of eight to twelve stems, tied together and placed in a paper bag, upside down in a warm, airy dry room. When the leaves are brittle, rub the sides of the bag together to remove the leaves from the stems. Package the dried basil and store in a cool, dark, dry place.

To tray dry, remove the leaves from the stalks, place them one layer deep on a tray and dry indoors, away from any direct sunlight, in a warm, dry, dark room that is well ventilated. The temperature should not exceed 100°F for the essential oils will be destroyed.

Bay Leaves

The aromatic leaves of the bay tree are used to season fish, game, poultry, stew, soups, salads, pickles, sauces, and almost any cooked vegetable.

Harvest young leaves when the tree is growing, dry them on a tray in a cool, well-ventilated room. Package in airtight containers and store in a dark, cool, dry place.

Beans, Fava

The method used to preserve fava beans depends on whether they are dried on the vine, or dried under controlled heat.

If the beans are dried on the vine, they must be treated so that any insects will be killed. Spread the dry, shelled, fava beans in a shallow pan to a depth of 1/2 to 3/4-inch and place in a 120°F to 145°F oven for three to four hours. If you plan to use the seeds for planting, keep the temperature below 135°F. Package the fava beans in an airtight container and store in a dry place or the beans may pick up insects once again.

For fava beans that are not dried on the vine, use mature beans that are tender. After they are shelled and washed, blanch them in boiling water for five minutes, drain and spread on the trays. Preheat the oven to 115°F, put the beans in, and gradually raise the temperature to 140°F. Stir the beans often. They should be dry in six to ten hours. Package the fava beans in an airtight container and store in a cool, dry, dark place.

Beans, Green or Wax

To prepare green beans, remove the ends and string if necessary. Cut smaller beans into 1-inch lengths, cut larger beans in half lengthwise. Try to make the pieces of uniform size to facilitate uniform drying. Steam the beans for 15 to 20 minutes, until tender but firm. Do not overcook. Spread the pretreated beans on trays to a depth of about 1/2-inch.

In an oven, begin with a temperature of 120°F–130°F, the lower temperature for whole beans, the higher temperature for beans that have been split lengthwise. Raise the temperature to 150°F after the first hour and lower it to 130°F when the beans are almost dry. Beans will dry in the oven in eight to ten hours. When dry, they are brittle to the touch and greenish-brown to greenish-black in color.

Oven dried beans should be conditioned and pasteurized before storing. Sun-dried beans may not need conditioning, but should be pasteurized.

Beans can also be dried in a room. Pretreat them as described above, but do not split lengthwise. Run a clean string through the upper third of each vegetable, keeping the beans about 1/2-inch apart. Hang the beans on their strings in a warm, dry, well-ventilated room until dry. Then pasteurize and store.

To rehydrate, soak one cup of beans in 2 1/2 cups of water for 2 1/2 hours.

Beans, Lima

Select lima beans that are mature and fully grown—beyond the stage where you would use them fresh or for canning or freezing. You can also leave them on the vine until the vines dry and then treat them immediately for storage.

Shell the beans and steam them for 10 to 20 minutes until they are tender but firm. You can also pretreat lima beans by immersing them in boiling water for five minutes, then draining them.

Spread the lima beans on drying trays to a depth of about 1/2-inch. In an oven, begin with a temperature of 120°F and gradually increase to 150°F. Stir frequently at the beginning. As the beans are nearly dry, lower the temperature to 140°F. Lima beans will dry in the oven in six to ten hours. When finished, they will be hard, brittle, break clean when broken, and shatter when hit with a hammer. Because they are low in acid, lima beans are not well-suited for sun-drying. They should be conditioned and pasteurized before storing.

To rehydrate, soak one cup of dried lima beans in 2 1/2 cups of water for 1 1/2 hours.

If the beans are dried on the vine, treat them so that any insects are killed. Spread the dried, shelled beans in a shallow pan to a depth of 1/2 to 3/4-inch and place in a 120°F to 145°F oven for three to four hours. Package the lima beans in an airtight container and store in a dry place or they may become infested with insects.

Beans, Navy

Navy beans can either be dried on the vine and then treated, or can be dried in the mature stage.

For navy beans that have been dried on the vine, treat them so that any insects are killed. Spread the dry, shelled navy beans in a shallow pan to a depth of 1/2 to 3/4-inch and place in a 120°F to 145°F oven for three to four hours. Package the navy beans in an airtight container and store in a dry place or the beans may pick up insects.

To oven-dry the beans, pick them when they are nearly full grown but before the pods are yellow and dry. Shell the beans and blanch them in boiling water for three minutes. Drain and spread them on trays to a depth of about one inch. Begin the drying at a temperature between 115°F and 120°F, then raise the temperature to 140°F. Stir frequently, especially at the beginning. Drying is complete when the navy beans are dry and brittle and show no moisture at the center when cut open.

Beans, Shell

Shell beans are often known by such names as pinto, kidney, and French horticultural. They can be dried on the vine and then treated, or picked at their peak and dried in controlled heat or in thc sun.

When dried on the vine, treat the beans so that any insects—especially weevils—are killed. Spread the dried, shelled beans in a shallow pan to a depth of 1/2 to 3/4-inch and place in a 120°F to 145°F oven for 3–4 hours. Package the shell beans in an airtight container and store in a dry place or the beans may become infested with insects.

If the beans are picked when they are mature, shell them, blanch in boiling water for five minutes and drain. They can also be steamed in shallow layers for ten minutes. Spread them on drying trays to a depth of no more than one inch. In controlled heat drying, there are two temperature ranges to choose from. Either begin at 115°F and raise the temperature to 140° for the rest of the drying, or begin at 140°F, raise the temperature to 160°F, then lower it to 130°F when the beans are almost dry. In either case stir the beans frequently. They will be brittle and hard when dry. Condition, pasteurize and store the beans in a cool, dry, dark place.

Sun-drying is not as effective as oven-drying for shell beans. If you do sun-dry, prepare the beans in same way as for controlled heat drying, condition if need be, and definitely pasteurize them. Store them in the same manner as oven-dried beans.

Beans, Snap

Trim and slice snap beans lengthwise or cut them into one-inch pieces. To pretreat, steam the beans for 15 to 20 minutes or until tender but firm. Spread the beans about 1/2-inch deep on the trays.

For oven-drying, begin at a temperature of 130°F, gradually raising the temperature to 150°F and then lowering the temperature to 130°F for the last hour when the beans are nearly dry to prevent scorching. Pretreat snap beans in the same way for sun-drying. In either case, condition and pasteurize the dried beans, package them and store in a cool, dry, dark place.

Snap beans will take approximately eight to ten hours to dry. When dry, they will be brittle and have a dark green to brownish color.

To rehydrate, soak one cup of dried snap beans in 2 1/2 cups of water for 2 1/2 hours.

Beechnuts

Beechnuts are small triangular-shaped nuts that are the fruit of American and European beech trees. The nuts fall from the husks after the first frost and should be gathered quickly before the squirrels store them away for the winter.

Dry beechnuts several days in a cool, dry place, then package in an airtight container and store in a cool, fairly humid place.

Beets

For drying, select small tender beets with good color and flavor, free from woodiness. Wash them and cut off all but 1/2-inch of the crowns (to prevent the beets from bleeding during pretreatment).

To pretreat beets, steam them whole for 30 to 60 minutes, depending on their size, until they are cooked through. Cool them, then trim off the roots and crown. Peel and cut in one of the following ways: 1/4-inch cubes, 1/8-inch crosswise slices, shoestring strips, or coarse shreds. The finer the pieces, the more rapidly they will dry. Shreds are recommended for sun-drying because of speed.

Spread the pieces no more than 1/4-inch deep on the drying trays. Begin drying with a low temperature between 120°F and 130°F. Use 120°F for slices and the higher temperature for smaller pieces. Increase the temperature after an hour to between 140°F and 155°F. Decrease the temperature as the beets come close to being dry. At an average drying temperature of 155°F, beets will dry in 10 to 12 hours. When dry they will feel tough, leathery, or brittle. Condition if need be and definitely pasteurize, especially sun-dried beets. Package and store in a cool, dry place.

To rehydrate, soak one cup of dried beets in 2 3/4 cups water for 90 minutes.

Beet Tops or Greens

To be satisfactory, dried beet tops or greens must be used within several weeks of drying. Wash the leaves, remove the mid-rib, and steam the leaves for five to seven minutes. Spread the leaves on a tray no more than 1-inch deep. Dry in controlled heat at a temperature between 125°F and 140°F. Stir often.

Berries

Dried berries are not a satisfactory substitute for the fresh, canned, or frozen varieties. Strawberries or gooseberries, in particular, are not good prospects for drying since the distinctive color and flavor of these fruits is dissipated by heat.

Pick over the berries and discard the defective ones. Wash them. You can remove the leaves and stems at this point or wait until after the berries are dry, when it is easier. No other pretreatment is absolutely necessary. However, if you desire, you may *either* steam the berries for 30 to 60 seconds, *or* crack tougher skins by dipping the berries in boiling water for 15 to 30 seconds then rapidly immersing them in cold water. After either treatment, carefully pat the berries dry with a towel.

Spread the berries no more than two deep in drying trays lined with cheesecloth (to keep them from sticking). Begin drying with a temperature of 120°F–135°F for the first hour, then increase to 150°F. When the berries are two-thirds dry, decrease the temperature to 140°F.

Berries are dry when they rattle on the tray, do not show any moisutre when crushed, and are quite brittle. The drying time is between four to eight hours.

To rehydrate, soak one cup of dried berries in a cup of water for one or two hours.

Blueberries

Blueberries taste best preserved by canning or freezing. If you wish to dry blueberries, first crack their skins by dipping them in boiling water for 15 to 30 seconds, then immerse them rapidly in cold water, drain, and pat dry. Then arrange on drying trays in a single layer.

For controlled heat drying, begin at 120°F. After one hour raise the temperature to 130°F, increase to 140°F in another hour and continue at this temperature until the fruit is dry. Blueberries will rattle on the trays when they are almost dry; it may take as long as four hours to dry them. When dry, they are hard and do not show any moisture when crushed. Condition oven-dried blueberries before storing them.

If you wish to sun dry blueberries, prepare them in the same way as for oven drying. After they dry, pasteurize them and store in an airtight container.

Blueberry leaves may also be dried for use as a tea. Pluck the leaves from the stems, spread them on a tray with a mesh screen bottom and leave them to dry in a warm, airy room away from direct sunlight. When the leaves are dry, crush them slightly, package them in airtight plastic or glass containers and store in a cool, dark, dry place.

Borage

Candied borage flowers can be used to decorate cookies and cakes. Pick the flowers when they are still covered with dew and allow them to dry on paper towels. Beat one or two teaspoons of cold water into an egg white, causing as little foam as possible. Brush the flowers with the mixture, then dust them thoroughly with granulated sugar. Dry slowly in a warm room. Store in the refrigerator.

Dried borage flowers can also be used to make a tea, as discussed in the section immediately preceding this directory.

Broccoli

Prepare broccoli for oven or sun drying by trimming the pieces, then cutting them lengthwise into quarters or eighths to make them as uniform as possible. Steam the stalks for 8 to 15 minutes depending on their thickness.

In an oven, begin with a low temperature around 120°F and raise it after one hour to between 140°F and 150°F.

When dry, broccoli will be brittle. Condition if necessary, and pasteurize. Then cool the pieces and store them in a dark, dry, cool place.

Buffalo Berries

Shepherdia argentea is a hardy shrub that grows in the northern sections of central North America—especially in the Dakota badlands. The shrub yields an egg-shaped yellow or red fruit about 1/3-inch long that is prized for jellies but can also be dried, as Indians in the northern plains states used to do.

Spread the berries on drying trays on a sunny day, or put them in the oven, beginning at a temperature of 120°F and gradually increasing to 140°F. When drying is nearly completed, reduce the temperature to 120°F to prevent scorching. Pack and store in an airtight container.

The berries may be rehydrated by soaking one cup of berries in one cup of water for four to six hours. Cook with sugar to taste and serve with meat, like a cranberry sauce.

Burnet

Dry the leaves, stems, and seed on a tray in a warm, dry, well-ventilated room away from direct sunlight. The mixture can be used to flavor vinegar. Crush the dried mixture well with a mortar and pestle. Pour one quart of white vinegar over half an ounce of the pounded mixture. Store for ten days, shaking well once a day. Strain and store in an airtight bottle.

Dried burnet leaves can also be used to make a tea, as discussed in the section preceding this directory.

Butternuts

The butternut is often called the white walnut and is treated and dried in the same way as the black walnut. See the entry on **Walnuts, Black**.

Cabbage

Cabbage is best kept in cold storage or made into sauerkraut. But it can also be dried for later use in soup.

Select a cabbage that is ready for table use. Trim it by removing the core and the outer leaves. If you have a vegetable grater, use the blade for cole slaw or slice it with a knife into shreds about 1/8-inch thick.

Cabbage is pretreated for both oven and sun drying by steaming it for five to six minutes or until it is wilted. Place the shreds evenly on the trays no more than 1/2-inch deep because cabbage and all other leaf vegetables have a tendency to pack down during drying.

In the oven, begin with a low temperature of 120°F and increase to an average of 145°F over the course of drying. Be especially careful near the end of the drying cycle because the leaves dry faster than the rib of the cabbage and will scorch if the temperature is too high. At an average temperature of 145°F, cabbage will take between six to eight hours to dry.

You may have to condition oven-dried cabbage; certainly pasteurize both oven and sun-dried cabbage. When the cabbage is dry it will feel tough, crisp, and brittle.

To rehydrate, soak one cup dried cabbage in three cups of water for one hour.

Caraway

In Shakespeare's *Henry IV*, Falstaff was invited to share a dish of caraways and a pippin apple. The herb has also been a favored seasoning in rye bread, sauerkraut, soups and stews, cookies and cake, and is the key flavoring ingredient in Kummel liqueur.

When the seeds are ripe, cut off the upper half of the plant, hang it upside down in a warm, airy room until the seeds dry, then thrash or shake them free. Store in an airtight bottle.

Carnations

While attractive as fresh flowers, carnations may also be candied. Use an amount of sugar equal to the weight of the flowers. Add an amount of water that is twice the volume of the sugar—for example, with one cup of sugar add two cups of water. Cook over low heat until a thin, light syrup is formed. Place the carnations in the syrup and stir until all of the syrup is absorbed in the carnations. Roll the carnations in granulated sugar, allow them to dry thoroughly, and store in a cool, dry place.

Carob

The carob is a large evergreen tree which, in the United States, is grown mostly in southern California. The tree, native to the Mediterranean countries, bears foot-long, brown seedy pods that contain 40 percent sugar and six percent protein. The carob pods are also known as St. John's bread, after John the Baptist who called them "wild locusts."

The carob pods fall from the tree in the late autumn, those that remain can be knocked down with a long pole. Spread them in the sun for a few days until dry. Packaged in a rodent proof container, they will store indefinitely.

Carob is ground into a powder and used as a chocolate substitute in milk, candy, bread, waffles, hot cakes, etc. It is especially popular with people who are allergic to, or prefer not to have, chocolate in their diet.

Carrots

Carrots can be preserved so well in cold storage that they should only be dried if such storage is impossible, or if you wish to use the vegetables in a soup mixture.

Select crisp, tender young carrots free from woodiness. They will take longer to dry than more mature carrots, but the product will be tastier. Don't use carrots that have remained in the ground until the end of the growing season—they will be too woody and will yield an unsatisfactory dried product.

Wash and scrub the carrots with a stiff brush. There is no need to peel them, although you may if you desire. Leave some of the top on and steam whole carrots about 20 minutes until tender but firm. Then slice them into 1/8-inch rings, 1/8-inch strips, or 1/4-inch cubes, or shred them. You may also dice, slice, or shred the carrots before steaming and steam for eight to ten minutes, until tender but firm.

Spread the carrots not more than 1/2-inch deep on drying trays and put them in the oven at 130°F, raising the temperature to 150°F after one hour. Reduce the temperature when drying is nearly complete to prevent scorching. Carrots require between seven to nine hours to dry. They will be tough and leathery when dry and have a deep orange color.

To rehydrate, soak one cup of dried carrots in 2 1/4 cups of water for an hour.

Celery

Both the leaves and stalks of the celery plant can be dried. Dry the leaves as you would an herb and crumble into pieces when dried. Store in airtight containers.

To dry the stalks, prepare them by cutting them into 1/4–1/2-inch pieces. The pieces should be uniform in size to facilitate uniform drying. Steam for four to ten minutes until tender.

In the oven, begin at 130°F, raise the temperature to 150°F after an hour, then lower the temperature to 130°F again when the celery is nearly dry. When dry, celery is tough and brittle. Condition and pasteurize.

The procedure is the same for sun drying.

Cherries

The best varieties of cherries for drying are Tartarian, Bing, Lambert, Dikeman, and sour or pie varieties.

Sort the cherries, wash, and either leave them whole or remove the stems and pits. Crack the skins of whole cherries by dipping them in boiling water for 15 to 30 seconds. Cool them immediately and pat dry. Whole cherries may also be syrup-blanched.

If the cherries have been pitted, allow them to dry for about an hour. They need no further treatment. White cherries may be sulfured for 10 to 15 minutes if they are to be sun-dried.

For oven drying, spread the cherries one layer deep

on the drying trays. Place them in the oven at a temperature of 120°F and gradually increase the temperature to 145°F. Reduce the temperature in the last hour when the cherries are nearly dry to 130°F to prevent scorching. When dry, cherries will be leathery and slightly sticky. Drying time averages about six hours using controlled heat.

For sun-drying, pit the cherries and sulfur them for 15 to 20 minutes before drying. Both oven and sun-dried cherries should be pasteurized before storing.

To rehydrate, soak one cup of dried cherries in one cup of water for four to six hours.

Chervil

Chervil leaves look like a relative of parsley and taste like a combination of parsley and fennel. The herb has a wide variety of uses as a seasoning.

Cut the chervil leaves when the plant is seven or eight inches tall to get more than one crop per growing season. Spread the fresh leaves on cheesecloth-lined drying trays and place them in well-ventilated, warm, dry rooms away from direct sunlight.

When the leaves are dry, bottle them in airtight containers and store in a dark, dry, cool place.

Chestnuts

When chestnuts are mature, they fall to the ground and should be picked up immediately or they will lose quality very rapidly. Chestnuts can be dried by simply leaving them in their shells in a dry room. A faster method is to cut an X on the flat side, boil in water for ten minutes, remove the shell and skin, then dry in a 150°F oven until the nuts are hard and shriveled.

Store in airtight containers in a dry, cool place.

Chicory

Belgian or French endive, also known as witloof chicory, is a sharp-flavored crisp green used in salads. Magdeburgh chicory, to which we refer here, is a long, tapered root used to flavor coffee. It is especially popular in New Orleans.

Wash and scrape the roots after harvesting them in the fall. Slice them lengthwise and blanch the slices for five minutes in boiling water. Roast the chicory slices in an oven at 300°F. When dry and brittle, put the chicory through a grinder. Pack in a tightly sealed jar and store in a dark, cool place.

Coconut

Probably the easiest food to dry, coconut also makes a tasty dried appetizer.

Split the coconut and remove the meat. Using a vegetable parer, cut the coconut meat in thin slices, put it on a tray and leave it in a 200°F oven until dry. In this form it can be used in puddings, candy, or eaten as a snack.

If you add salt before putting the coconut slices in the oven, you can use the dried coconut as an appetizer.

To dry coconut for grating, split the shell and dry the meat in the oven or sun until it easily separates from the shell. Remove the shell and continue drying the meat in the oven until it is crisp, hard, and has a uniformly off-white color. In this form, you can grate the coconut, package it in airtight containers and store it at room temperature for several months.

Coriander

Coriander's sweet-tasting seeds are used to flavor baked goods, salads, sauces, curries and liqueurs.

Harvest the seeds when the plants are mature. Cut off the top of the plants, gather them into bunches, and hang them upside down in a warm, well-ventilated, dry room away from any direct sunlight.

When dry, shake the seeds free, bottle and store in a dark, dry, cool place.

Corn, Sweet

Select any good table variety of corn in the milk stage—that is, just right for eating as roasting ears. It is important to start drying the corn immediately after gathering it.

Husk the corn and sort the ears according to maturity. It is not necessary to silk the corn. Young corn requires more steaming time to set the milk than does older corn.

Steam the corn on the cob or immerse in boiling water until the milk is set—about 10 to 15 minutes for older corn, 15 to 20 minutes for younger corn. Steaming or blanching is finished when no fluid escapes from kernels that have been cut across.

When the corn has cooled, cut the kernels from the cob, making sure that no cob is removed with the kernels. Kernels can be cut at about two-thirds of their depth.

Spread the kernels on the drying trays to a depth of 1/2 to 3/4-inch.

For oven drying, begin at a temperature of 130°F gradually increasing the temperature to 165°F, reducing the temperature to 140°F for the last hour when the corn is nearly dry. If the temperature is held at 165°F for at least an hour it is not necessary to pasteurize oven-dried corn. Stir the corn frequently to separate the grains and break up any masses.

For sun drying, corn is prepared and pretreated in the same way as for oven drying. Sun-dried corn must be conditioned and pasteurized before storing. When oven-dried corn is conditioned, the silk is easier to remove. Corn is dry when it is hard, brittle, somewhat transparent, breaks clean, and shatters when hit with a hammer. Be careful throughout the drying process for it scorches easily.

Corn shrinks rapidly during drying and the contents of two or three trays can be combined after the shrinkage has taken place. Corn requires an average of six to ten hours to dry.

To rehydrate, soak one cup of dried corn in 2 1/4 cups of water for 30 minutes.

For variety, you might want to try the cream method of drying corn. Clean, cook and cut the corn as described above. For each gallon of corn, add 1/2 cup of salt, 1/2 cup of sugar and one cup of sweet cream. Mix thoroughly and dry in the oven. Corn dried in this manner should be dried in one day, conditioned for 24 hours, and immediately sealed in airtight containers. The corn must be dried quickly to prevent mold and stored in a cool, dark place to prevent it from becoming rancid. Rehydrate the same as above.

A corn snack can be prepared from sweet corn by pulling the corn husks back just enough to remove the

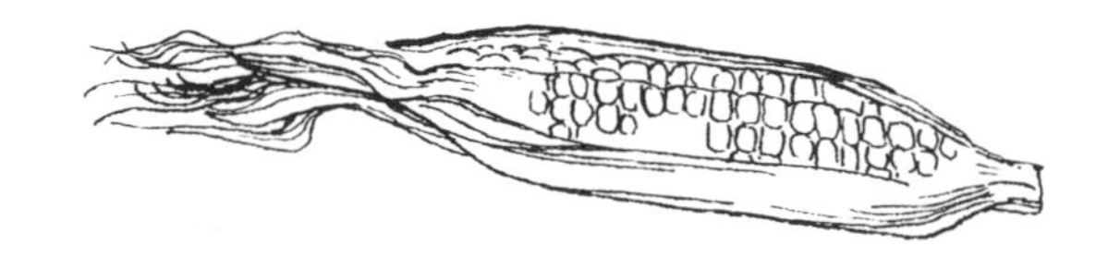

silk. Close the husks over the ears again and roast the ears over hot coals or in a hot oven (450°F.), turning frequently, for 25 minutes. Remove from the heat, air-cool and strip the kernels from the cobs. Spread the kernels in a shallow tray and oven-dry at 165°F. for one hour or until corn becomes brittle. Brush coconut oil over the corn and salt it during the last 15 to 20 minutes of drying. The drying can also be completed in the sun. This takes three days and the trays should be brought inside at night.

Cowpeas (Black-eyed Peas)

The actual name for southern black-eyed peas is cowpeas. They can be dried on the vine or under controlled heat.

If the cowpeas are dried on the vine, they must be treated to kill any insects. Spread the dry, shelled peas in a shallow pan to a depth of 1/2 to 3/4-inch and place in a 120°F to 145°F oven for three to four hours. If the cowpeas are to be used for planting, keep the temperature below 135°F. Package in an airtight container and store in a dry place or the cowpeas may become infested with insects.

For cowpeas that are not dried on the vine, use green, plump cowpeas. Shell, wash, and blanch in boiling water for three minutes, drain, and spread on drying trays in shallow layers. Place the trays in an oven preheated to 115°F and gradually raise the temperature to 140°F. Stir the cowpeas often. They should be dry in six to ten hours. Package in airtight containers and store in a cool, dry, dark place.

Cranberries

Crack the skins of the cranberries by dipping them in

rapidly boiling water for 15 to 30 seconds, then quickly immerse them in cold water. Towel dry.

For oven drying, place the cranberries on drying trays one layer deep. Start the oven at 130°F, then raise the temperature to 140°F after an hour or two. Cranberries will rattle on the trays when dry, will be hard, and will show no moisture when crushed. Condition, pasteurize and store.

The procedure is the same for sun drying.

To cook, pour boiling water over to cover and simmer for 10 to 15 minutes. Sweeten after cooking. To use in a pie, pour water over and let the cranberries stand for three to four hours.

Cumin

Cumin, one of the herbs mentioned in the Bible, has a long history as a seasoning ingredient. Today, cumin seeds are used for pickles, cheese, bread, sauerkraut, and chili and curry powders.

When the seeds are dry, harvest the entire plant and shake out the seeds. Package and store in an airtight container.

Currants

Crack the skins of the currants by dipping them first in rapidly boiling water for 15 to 30 seconds. Then quickly immerse them in cold water. Pat dry.

For oven drying, spread the currants on drying trays one layer deep. Place in the oven at 130°F, then raise the temperature to 140° F after an hour or two. Currants will rattle on the trays when dry, will be hard, and show no moisture when crushed. Condition, pasteurize, and store.

Drying in the sun utilizes the same procedure.

Dates

Choose dates with low moisture content. Translucent dates are the best for drying. Wipe the fruit clean with a damp terrycloth towel. Do not wash. Discard any dates with damaged skins, fungus, or any signs of souring. Dry only one kind of date at any one time. Different varieties have different moisture contents and will dry at different rates.

To oven dry, spread the fruit one layer deep on drying racks or trays. Dates need no pretreatment. The drying procedure for dates is different from any other fruit. Preheat the oven to 225°F for five minutes. Turn the heat off and place the dates in the oven until the oven cools. Repeat the process the next day and as many times as are necessary until the dates are dry and ready to be stored, which may take several days.

Another oven method is to place a thermometer in one of the dates. Turn the oven on until the thermometer reads between 120°F and 125°F. Then turn the oven off until the thermometer reads between 105°F and 110°F. Repeat the process as many times as necessary until the dates are dry and ready to store. Dry dates until they are leathery, pliable, and slightly sticky. Never allow the temperature of the fruit to go above 155°F or the sugar will caramelize, giving the dates a scorched flavor and making them syrupy and sticky.

Dates can also be dried in a sun dryer but will take between two to eight days to dry completely, depending on the climate and the moisture level of the fruit.

Oven drying will effectively kill any insects and insect larvae on dates. To accomplish the same end with sun-dried dates, simply place them in the oven—before or after drying—at 150°F for 30 minutes.

Dried in the oven until chewy and firm and packed in airtight containers, semi-dry varieties of dates such as

Medjool, Zahedi, and Deglet Noor will keep for several years at room temperature. Soft varieties do not keep nearly as long or as well. Use them within six months because their flavor deteriorates over longer periods in storage.

Dill

Dill was used during the Middle Ages to cast spells against witches. Today, the herb is a popular food seasoning in many parts of the world. Both seeds and leaves are used, but the flavor is most delicate in the latter.

To dry dill leaves, collect them just before the flowers open. A bunch can be tied together and suspended upside down in a warm, airy room away from direct sunlight. The leaves can also be dried in the oven. Use a temperature no higher than 140°F or you may lose the essential, aromatic oils that give dill its flavor. You can also dry the leaves by using oven heat remaining after baking. Store the leaves in airtight containers in a dark, dry place.

Dill seeds are collected from the dried flower heads and stored immediately in the same manner as the dill leaves.

Elderberries

Elderberries are dried in much the same way as blueberries. Begin by cracking their skins, plunging them in boiling water for 15 to 30 seconds, then immersing them in cold water. Drain and towel dry.

Spread the berries on drying trays in a single layer. In controlled heat drying, begin at a temperature of 120°F. Raise the temperature to 130°F after an hour, increase again to 140°F in another hour and continue at this temperature until the fruit is dry. Elderberries will rattle on the trays when they are almost dry, which should take about four hours in controlled heat. Dry berries will be hard and show no moisture when crushed. Condition oven-dried elderberries before storing them.

Treat elderberries in the same way for sun drying. After they dry, pasteurize them and store in airtight containers in a cool, dry place.

The blossoms or flowers of the elderberry may also be dried for use as a tea. Pick them when they are in full bloom, remove the stems, spread the flowers on a tray with a mesh screen bottom, and leave them to dry in a warm, airy room away from direct sunlight. When they are dry, package them in an airtight plastic or glass container and store in a cool, dark, dry place.

Fennel

Dry the fennel leaves on a tray in a warm, airy, dry room away from any direct sunlight. Store in airtight, small containers.

The seeds are harvested from the flower heads when dry and stored immediately for use in fish, meats, salads, soups, cheese and egg dishes, and baking.

Figs

The best varieties of figs for drying are Adriatic, Black Mission, Calimyra, Brown Turkey, Kadota, Smyrna Celeste. Select fully ripe fruit, or the sugar content will be too low to produce a good dried product and the fruit may sour. Figs left on the tree are ready for drying when they fall to the ground.

Small figs can be left whole. Larger, juicier figs should be cut in half lengthwise.

To pretreat figs, choose one of the following methods. Figs left whole should have their skins cracked by dipping them in boiling water for 30 to 45 seconds, cool quickly, and remove all excess moisture. Light-colored varieties, such as Kadota, should be sulfured for two to three hours. Sulfuring is optional with other varieties except black figs, which should not be sulfured because they turn an unappetizing mottled color. If figs are cut, steam them for 20 minutes. Or, they may be syrup-blanched for a sweeter product.

To oven dry, begin at a temperature of 115°F. Gradually increase the temperature to 145°F after the first hour and reduce to 130°F when the figs are nearly dry to prevent scorching. Stir or turn figs to keep them from sticking. When dry, figs are leathery, glossy, with dry skin and flesh that is pliable yet slightly sticky. Drying time in the oven is about five hours for halves, 10 to 12 hours for whole figs. Sun drying takes about three days.

Preparation and pretreatment for sun drying is the same. Condition oven dried figs, pasteurize sun-dried figs, package and store in a cool, dark place.

Filberts (Hazelnuts)

Harvest filberts when the nuts have fallen to the ground. It is not important to gather each nut as it falls, but the nuts should not be left on the ground too long because they will begin to discolor.

To sort the good nuts from the bad, put them in a tub of water and discard those that float. You can shell them or not, but shelled nuts will dry faster.

Spread the nuts on drying trays a few layers deep. The optimum drying temperature for filberts is between 95°F and 105°F. If the temperature goes above 110°F the quality of the nuts will be diminished. Air circulation is important to get the nuts dried as quickly as possible. An electric heater with a fan may be the best way to dry filberts. Drying time is approximately two days.

Filberts are firm at the start, become spongy while drying, and firm up again when dry. The internal color gradually changes from white to a creamy color beginning from the outside. When the color changes in the center of the kernel, the nut is dry. A cold, dried kernel will snap when bitten.

Garlic

Garlic is seldom dried because it keeps for approximately eight months if conditioned after harvesting.

If you want to try drying garlic, however, harvest it when it is ready for table use—when the tops fall over. Slice in uniform slices at least 1/8-inch thick. Anything thinner will probably be scorched. Pretreatment by steaming is not necessary, although you may steam the slices for 30 to 60 seconds if you desire.

Place the garlic slices on the drying trays. Put the trays in a 135°F–140°F oven, lowering the temperature to 130°F when the slices are nearly dry to prevent scorching. Garlic is light-colored and brittle when dry. Condition and store.

The sun drying procedure is the same as for oven drying, but sun-dried garlic slices must be pasteurized before storing.

Dried garlic may be crushed or ground for use as a garlic powder seasoning.

To rehydrate, soak one cup of dried garlic in two cups of water for 45 minutes.

Ginger

Ginger, an important ingredient in oriental cooking, can be dried, candied, or crystallized. Ginger is grown outdoors in warm climates as well as indoors in greenhouses.

Harvest the roots in early winter and place in a well-ventilated, dry room until completely dry. The ginger must then be finely ground after the skin is removed. Store ground ginger in an airtight container and use as a seasoning.

To candy ginger, remove the skin from young roots and cut into 1/8-inch thick slices. Boil in water to cover for five minutes, drain and repeat three more times. Save the liquid from the last boiling, measure the ginger and liquid, and add 1 1/2 times as much sugar. Boil until the ginger is clear.

For candied ginger with a syrupy consistency, pour the ginger and syrup into a hot, sterilized jar and seal immediately. For crystallized ginger, drain the ginger as completely as possible after boiling in the syrup and roll in granulated sugar. Pack in airtight glass jars.

Grapes (Raisins)

For drying, use any variety of seedless grape—Thompson will produce white raisins and Monukka (sometimes called Black Thompson) will produce large, black raisins.

Wash the grapes and remove all defective fruit. Either leave the grapes on the stem, cutting tight bunches into smaller batches, or remove all of the grapes from the stems. No pretreatment is necessary, but it's advisable to crack the skins of any especially thick-skinned batches by dipping them in rapidly boiling water for 15 to 30 seconds, then plunging them into cold water immediately to cool. Drain the grapes after cracking and towel dry.

If you stem the grapes, spread them one layer deep on the drying trays. If you leave them on the stems, be sure to turn the batches throughout the drying process to assure even drying.

In controlled heat, begin at a temperature of 120°F, gradually increase to 150°F, then reduce to about 130°F near the end to prevent scorching. Drying time may be as much as eight hours. Dry grapes until they are pliable and leathery—just like commercially-produced raisins.

Follow the same procedure for sun drying grapes. The drying process, however, may take as long as two weeks in a hot dry climate. Be careful not to overdry or scorch the grapes in the sun. Pasteurize after drying and pack the raisins for storage in airtight containers to prevent insect infestations.

Hickory Nuts

Hickory nuts are the fruit of the shagbark and shellbark hickory trees, members of the walnut family. Hickory trees have also been crossed with pecans to produce a nut called the hiccan. The hickory nut is mature when it falls to the ground.

Dry the nuts in the shade for several days, then pack in airtight containers and store in a cool, dark, dry place.

Huckleberries

Huckleberries are best preserved by canning or freezing. If you wish to dry huckleberries, first crack their skins by dipping them in rapidly boiling water for 15 to 30 seconds, then plunging them immediately into cold water to cool them. Drain the berries after cracking the skins and towel dry.

Spread the berries on drying trays one layer deep. For oven drying, begin at 120°F, raise the temperature to 130°F after one hour and then raise it again in another hour to 140°F. Huckleberries will be hard and rattle on the trays when they are dry and show no moisture when crushed.

Prepare huckleberries the same way for sun drying. After they are dry, pasteurize and store in a cool, dry place.

Jonquils (Daffodils)

Jonquil flowers may be candied. Cut the stalk within 1/4-inch of the flower. Make a thin syrup of one part sugar to two parts water and heat it on the stove. Place the jonquil flowers in the syrup, stirring them in lightly and taking care that the flower retains its form but is fully covered with syrup. Sift granulated sugar over the flowers after they are removed from the syrup. Dry on a rack, package in airtight jars and store in a dry place.

Jujube

The fruit of the jujube tree is not the candy of the same name, but rather an oblong brown fruit about one inch long. Jujube trees, originally native to China, are grown in the warmer parts of the United States, especially the southwestern states and southern California. Because the jujube bears a resemblance to dates it is sometimes called the Chinese date.

Jujubes can be dried or candied. The best way to dry jujubes is leave the fruits on the trees for a few days after they turn dark brown—allowing them to dry naturally in the sun. Then harvest and store in airtight containers placed in a dry, dark, cool place.

To candy jujubes, harvest them when they turn red-dish-brown. Score the fruit with a knife or razor blade and treat in the manner prescribed for candying fruit on page 43. Pack the dried candied fruit in shallow containers, placing a layer of waxed paper between each layer of candied jujubes. Cover tightly and store in a cool, dry place.

Kale

To be satisfactory, dried kale must be used within several weeks of drying. Wash the leaves, remove the midrib, and steam the leaves for five to seven minutes. Spread the leaves on a tray no more than one-inch deep. Dry in an oven at a temperature between 125°F and 140°F. Stir often to prevent scorching.

To cook, simply plunge the dried greens into boiling water. No prior rehydration is needed.

Lemon Balm (Melissa)

Cut the entire lemon balm plant in midsummer before the flowers appear. Dry the leaves by tying bunches together and suspending them upside down in a warm, dry, well-ventilated room away from any direct sunlight. Shake the leaves from the stems when dry, package in glass or plastic containers and store in a cool dry place.

The dried leaves, with a taste reminiscent of lemon, can be brewed into an excellent tea, either by themselves or in combination with other herbs such as lemon verbena and mint. The leaves are also used to season broiled fish, pickled herring and chowders.

Lentil

The lentil was named after the object it resembles—a lens. Since lentils are generally dried on the vine, they must be treated to kill any insects.

Spread the dry, shelled lentils in a shallow pan to a depth of 1/2 to 3/4-inch and place in a 120°F to 145°F oven for 3 or 4 hours. Package in an airtight container and store in a dry place or the lentils may become infested with insects.

Lilacs

Lilacs can be crystallized in a sugar and gum arabic mixture. To make gum arabic, dissolve one cup of sugar in 1/2 cup of water and cook over low heat until clear. Cool the syrup slightly and thoroughly blend in four teaspoons of powdered acacia. Chill the gum arabic before using.

To crystallize lilacs, dissolve one ounce of the gum arabic mixture in 1/2 cup of hot water and let cool. Dip small bunches of lilacs in the mixture and let them dry while mixing together 1/2 cup of water, one cup of sugar, and one tablespoon of corn syrup and cooking until it reaches the softball stage (234°F to 238°F). Dip the lilacs into the mixture, remove and sprinkle them with granulated sugar. After drying, store in a cool, dry place.

Macadamia Nuts

The macadamia tree, originally from Australia, now bears nuts in California, Florida, and Hawaii. When macadamia nuts are mature, they fall to the ground and must be harvested immediately and frequently to prevent spoilage. Remove the husks and spread the nuts on drying trays in a shady area to dry for two to three weeks. Then pack the nuts in an airtight container and store in a cool, dark, dry place.

Marigolds

Marigold flowers can be dried for use in teas, as well as a seasoning in stews, soups, chowders, salads—even on a cheese or liverwurst sandwich. They can also be candied.

To dry the flowers, use only the florets. Dry them quickly on a drying tray in a warm, well-ventilated room away from direct sunlight. The florets should not touch one another or they will change color. When dry, remove the petals from the flowers, bottle, and store.

To candy marigold flowers, heat one part sugar and two parts water over a low flame until a thin, light syrup is formed. Place the marigolds in the syrup and stir gently until all of the syrup is absorbed in the marigolds. Roll the flowers in granulated sugar and, after they dry, store in a cool, dry place.

Marjoram, Sweet

The ancient Greeks and Romans planted marjoram on graves to insure the peace of their departed relatives and used sprigs to crown a newly married couple for a peaceful life. Today marjoram is a popular seasoning for meats, poultry, vegetables, stuffings and sausages.

Cut the leaves from the top of the plant several times during the growing season. Dry them on trays in a warm, well-ventilated, dry room away from any direct sunlight. Bottle and store in a cool, dark, dry place.

Mimosa

Select the flowers when in full bloom and dry them on a drying tray with a wire mesh screen in a warm, well-ventilated room away from direct sunlight.

The dried flowers may then be crystallized. Begin the process by making a gum arabic, dissolving one cup of sugar in 1/2 cup of water and cooking over low heat until clear. Cool the syrup slightly and thoroughly blend in four teaspoons of powdered acacia. Chill the gum arabic before using.

To crystallize mimosa, dissolve one ounce of the gum arabic mixture in a half cup of hot water and let cool. Take a cup of the dried flowers and dip them, one at a time, into this mixture. Let them dry while mixing one cup of sugar and one tablespoon of corn syrup into 1/2 cup of water, and cooking until it reaches the softball stage (234°F to 238°F). Dip the flowers into this mixture, remove, and sprinkle them with finely granulated sugar. After allowing them to dry thoroughly, package and store in a cool, dry place.

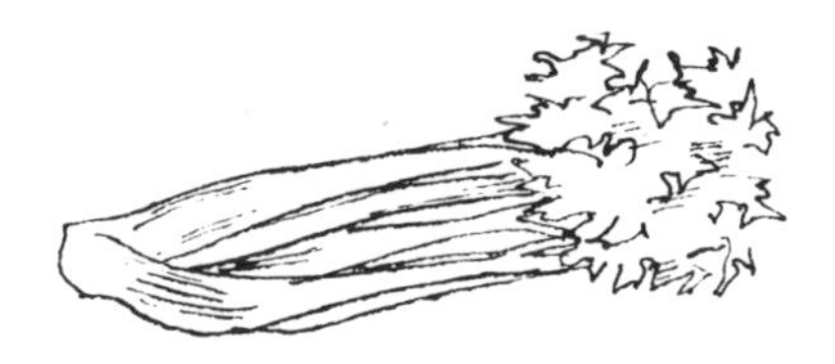

Mint

Collect mint early in the morning, just after the dew has evaporated. Several sprigs can be tied together and suspended upside down to dry in a warm, airy room away from direct sunlight. Shake the leaves from the sprigs when dry. Mint leaves removed from the sprigs may also be dried on trays in a warm, well-ventilated room. Or they may be dried in an oven, using a temperature no higher than 140°F to be certain not to lose the essential aromatic oils that give mint its delightful scent and taste. Store dried mint leaves in an airtight container in a cool, dark, dry place.

Mushrooms

Wild and cultivated mushrooms are both dried in the same manner, but you are well-advised to use only the cultivated varieties unless you are truly a mushroom expert. While few wild mushrooms are poisonous, there are enough to suggest that you should be an experienced mushroom hunter before eating or preserving wild mushrooms. There are no simple tests to enable you to distinguish between edible and poisonous mushrooms.

Select young, fresh mushrooms for drying. They should not be bruised or turning brown. If necessary, wipe them clean with a damp cloth. Larger mushrooms may be peeled and sliced; smaller ones can be dried whole. Remove the stems from larger mushrooms. If they are tender, slice the stems for drying; if tough, discard them.

Sliced caps and stems should be approximately 1/8-inch wide. To prevent discoloration, coat the mushrooms with ascorbic acid. Precooking is not necessary, but you may steam the mushrooms for 12 to 15 minutes if you wish.

To sun dry, spread the mushrooms in trays, or on paper or cloth not more than 1/2-inch deep. Condition the mushrooms after drying, and definitely pasteurize them.

In a dryer, maintain the temperature at 130°F for one hour, then raise the temperature slowly to 150°F. When the mushrooms are almost dry, reduce the temperature to 140°F until the drying is finished. Again, be sure to condition and pasteurize the mushrooms after drying. Mushrooms are dry when they are leathery or brittle.

After conditioning and pasteurizing, cool the mushrooms and pack them in a clean jar with a tight seal. Store in a cool, dry, dark place.

Mustard (Seeds And Greens)

The seeds of both the black and white mustard plants may be used for drying. Harvest the 1-inch seed pods when the lowest ones begin to dry on the plant. Place the unopened pods on trays or cheesecloth in a warm, dry place until they dry. Then shake out the seeds and bottle them. The seeds can be ground to make dry and prepared mustard.

Dried mustard greens must be used within several weeks of drying to be satisfactory. Wash the leaves, remove the mid-rib, and steam the leaves for five to seven minutes. Spread the leaves on a tray no more than one inch deep. Dry in an oven at a temperature between 125°F and 140°F. Stir often to prevent scorching.

Nectarines

Select fully-ripened fruit. Leave the skin on or remove it by blanching. Fruit with skins on will take longer to dry or sulfur. If the skin is left on, wash well.

Cut the nectarine in half and remove the pit. Either leave in halves, or cut into quarters or slices. Again, the smaller pieces will dry and sulfur faster. Coat with ascorbic acid solution while preparing.

Pretreat nectarines in one of the following methods. Steam blanch the halves for 15 to 18 minutes, slices for five to seven minutes, quarters for eight to ten minutes. Or sulfur, using two teaspoons of sulfur per pound of cut fruit. Sulfur slices for one hour, halves or quarters for two hours.

Arrange the fruit in a single layer on the drying trays,

pit side up for the halves. Be careful not to spill any juices that have accumulated in the halves during sulfuring.

For oven drying, begin with a temperature of 130°F, gradually increasing the temperature to 155°F, and reducing the temperature to 140°F during the last hour or when the nectarines are nearly dry. Do not turn the nectarine halves until the juices in the pit hollow have disappeared.

The average drying time for halves is 14 to 15 hours in controlled heat, six hours for slices, and two or more days for either in the sun. When dry, nectarines are leathery, pliable, soft, and show little or no moisture when cut in half.

Condition, pasteurize, package, and store.

To rehydrate, soak one cup of dried nectarines in two cups of water for six to eight hours.

Okra

Select tender young okra pods. Steam them whole for five to seven minutes and then cut into rings 1/2 to 1-inch long. Keep the slices uniform to facilitate even drying. Dry in an oven starting at 125°F, increasing the temperature to 145°F after one hour and reducing to 130°F as the okra is nearly dry to prevent scorching. Okra takes between six to eight hours to dry and when finished is brittle to the touch and green in color. Okra is not a good vegetable to sun dry.

To rehydrate, soak one cup of dried okra in three cups of water for approximately one hour.

Onions

Select the pungent varieties of onions for drying.

Peel off the outer leaves and cut the onions into uniform slices 1/8 to 1/4-inch thick. Uniform slices are very important since thin slices will brown or scorch. No pretreatment is needed for oven or sun-dried onions.

For oven drying, spread the onions in a shallow layer on the drying trays and place in an oven at a temperature between 135°F to 140°F. Maintain the same temperature throughout the drying process. Dry until the onions are brittle, crisp, and light-colored—approximately six to ten hours.

Prepare the same way for sun drying. Condition oven-dried onions; pasteurize sun-dried onions. Package and store.

Dried onions may be crushed for use as an onion powder seasoning.

To rehydrate, soak one cup of onions in two cups of water for 45 minutes.

Orange Flowers

Orange flowers may be dried for use as a flavoring or as a tea. Select flowers in full blossom and dry on trays in a warm well-ventilated room.

Fresh orange flowers may also be candied. Add one part sugar to two parts water, heat over a low flame until a thin, light syrup is formed. Place the orange flowers in the syrup and stir until all of the syrup is absorbed in them. Roll the orange flowers in granulated sugar, dry on a rack, and store in a cool, dry place.

Oregano

This herb is often called wild marjoram. Like many herbs, oregano may be dried by cutting sprigs from the plant before the flowers form, tying them together in small bunches and suspending them upside down in a warm, airy room. Shake the leaves from the sprigs after they've dried completely.

The leaves may also be cut from the plant, again before the flowers form, and dried on trays in a warm, dry, well-ventilated room away from any direct sunlight. Crush the leaves after drying, bottle and store in a cool, dark, dry place.

Pansies

Pansy petals can be crystallized in a sugar and gum arabic mixture. To make gum arabic, dissolve one cup of sugar in 1/2 cup of water and cook over a low heat until clear. Cool the syrup slightly and thoroughly blend in four teaspoons of powdered acacia. Chill the gum arabic before using.

To crystallize one cup of pansy petals, dip them in a mixture of one ounce of the gum arabic dissolved in 1/2 cup of hot water. Let the petals dry while mixing one cup of sugar and one tablespoon of corn syrup into 1/2 cup of water and cooking until it reaches the softball stage (234°F to 238°F). Dip the dried pansy petals in this mixture, remove and sprinkle with granulated sugar. Allow them to dry thoroughly, package and store in a cool, dry place.

Parsley

Rich in vitamins A, B, and C, parsley also contains calcium, copper, iron and manganese. It is one of the most nutritious of all the herbs—certainly more than just a garnish.

To dry parsley, cut young branches from the plant and wash them lightly under cold water to remove any dirt and dust. Shake dry and hang the parsley in bunches upside down in a warm, airy room away from direct sunlight until dry. Or spread the leaves out on a tray.

It is also possible to dry parsley in the oven. In a 400°F oven, parsley will dry in as little as five minutes, but vitamins that are affected by heat may be destroyed in the process. It is more advisable to dry the parsley in a 100°F to 125°F oven, making certain the temperature never rises above 140°F so that the plant's essential oils and vitamins are not adversely affected.

When dry, put the leaves through a coarse sieve, pack and store in a cool, dark, dry place.

Parsnips

Wash, trim, and peel parsnips to prepare them for pretreatment. Steam whole parsnips about 20 minutes until tender but firm. Then slice them crosswise to a thickness of about 1/8-inch, dice them into 1/4-inch cubes, or shred them. The parsnips may also be sliced, diced or shredded before steaming, in which case they should only be steamed for about ten minutes.

Spread the parsnips not more than 1/2-inch deep on drying trays and place in a 150°F oven. Reduce the temperature when drying is nearly completed to prevent scorching. When dry, parsnips will be very brittle.

To rehydrate, soak one cup of dried parsnips in 2 1/4 cups of water for one hour.

Peaches

The best varieties of peaches for drying are the freestones, such as Elbertas, but any good table variety will do. Select fully ripened fruit. Leave the skin on or remove it by blanching. Fruit with skins will take longer to dry or sulfur. If the skin is left on, wash the peaches well to remove the fuzz. Cut the peaches in halves and remove the pits and any red-colored parts in the hollow which would greatly discolor when the fruit is dried. Either leave in halves, or cut into quarters or slices. The smaller pieces will sulfur and dry faster. Coat with ascorbic acid solution while preparing to prevent discoloration.

Pretreat the peaches by one of the following methods. Either steam blanch the halves for 15 to 20 minutes, slices for five to seven minutes. Or sulfur, using two teaspoons of sulfur for every pound of cut fruit. Sulfur between one to two hours, depending on the size of the pieces.

Arrange the fruit in a single layer on the drying trays, pit side up for the halves. Be careful not to spill any juices that have accumulated in the halves during sulfuring.

For oven drying, begin with a temperature of 130°F, gradually increasing the temperature to 155°F, and reducing the temperature to 140°F during the last hour when the peaches are nearly dry. Turn the peaches over when the juices in the pit hollow have disappeared.

The average drying time for halves is 15 hours in controlled heat, about six hours for slices, and two or more days in the sun. When dry, peaches are leathery, pliable, soft, and show little or no moisture when cut in half.

Condition, pasteurize, package, and store.

To rehydrate, soak one cup of dried peaches in two cups of water for six to eight hours.

Peanuts

Select mature peanuts for drying. They require no preparation or pretreatment. Spread them on trays and dry in the sun, or put them in a 130°F oven, raising the temperature to 150°F after one hour.

The peanuts can either be shelled and skinned before storing, or stored with shells and skins still on.

Peanuts may be roasted at a temperature of 300°F for 30 to 45 minutes in their shells or 20 to 30 minutes if shelled. Stir frequently to prevent scorching.

Pears

The best variety of pears to dry is Bartlett, but any fine-grained pear with good flavor and high sugar content dries well. Pick the pears before they are ripe. Store them for approximately one week at a room temperature not over 70°F until ripe, but still firm.

Pare off the skin and split the fruit in half lengthwise. Remove the core and the woody stem and vein. Leave in halves, quarter, or cut into 1/8 to 1/4-inch slices. While preparing the pears, coat them with ascorbic acid solution, to prevent discoloration.

Pretreat pears in one of the following ways. Either steam them for five to twenty minutes depending on the size of the pieces. Or sulfur, using 2 1/2 teaspoons of sulfur for every pound of cut fruit. Sulfur for one or two hours depending on the size of the pieces. Or syrup blanch the fruit as described in the preceding section.

Spread the pears on drying trays not more than one layer deep. For oven drying, begin with a temperature of 130°F, gradually increasing the temperature to 150°F and reducing it to 140°F during the last hour or when the pears are nearly dry.

The pears will be springy and suede-like when dry, and no moisture will be evident when the fruit is cut or squeezed. The average drying time under controlled heat is 15 hours for halves or six hours for slices. The fruit will require at least two days to dry in the sun.

To rehydrate, soak one cup of dried pears in two cups of water for four to six hours.

Peas

Select young tender peas of a sweet variety. They can be picked at their peak or dried on the vine and then treated. If picked when young and tender, shell the peas and steam them for 15 to 20 minutes until tender but firm. Peas can also be pretreated by immersing them in boiling water for three minutes, draining and then toweling them dry.

Spread the peas on drying trays to a depth of about 1/2-inch. To oven dry, begin with the temperature at 120°F and gradually increase to 150°F. Stir frequently at the beginning. As the peas are nearly dry, lower the temperature to 140°F to prevent scorching. Peas will be dry in the oven in seven to nine hours. When finished, they will be hard, brittle, wrinkled, and shatter when hit with a hammer. Condition, pasteurize, package and store.

Sun drying is not as effective as oven drying for peas. If sun dried, prepare the same way as for oven drying. Condition if needed and definitely pasteurize the peas.

If the peas are dried on the vine, treat them so that any insects or larvae are killed, by spreading the dried, shelled peas in a shallow pan to a depth of 1/2 to 3/4-inches and place in a 120°F to 145°F oven for three to four hours. Package the peas in an airtight container and store in a cool, dry place or they may become infested with insects.

Pecans

Most pecan trees in the United States are wild, but they have been cultivated at least as far back as the time Thomas Jefferson brought several trees up from the Mississippi Valley to Virginia. He gave a few to George Washington and three are still growing at Mount Vernon.

The pecan nut is mature when the green husks turn brown and open. The nuts will fall to the ground or can be knocked down with long poles.

If the weather has been humid, spread the nuts out to dry in a warm, airy place. When dry, store in an airtight container. Pecans stored in a refrigerator or freezer will keep for as long as one year. At room temperature they will last approximately four to six months.

Peppers (Sweet, Green, Bell, Pimentos)

Peppers can be dried in strips, quarters or rings. To dry quarters or strips, cut the peppers in half and remove the seeds and cores. Strips should be uniform size, a minimum of 1/2-inch thick. To dry rings, cut the pepper into rings, again about 1/2-inch thick, and remove seeds and core.

Steam peppers for 10 to 12 minutes. Spread pepper rings and quarters two layers deep on the drying trays, slices not more than 1/2-inch deep.

For oven drying, begin at 120°F, gradually increasing the temperature to 150°F after one hour. Reduce to 130°F or 140°F when the peppers are nearly dry to prevent

scorching, using the lower temperature if the peppers are thin. Peppers are pliable and medium green in color when dry. Condition, package, and store.

For sun drying, follow the same procedure. Pasteurize, package and store.

Peppers, Green Chili

Select chili peppers that are full-grown and bright green in color. Wash the peppers. To remove the skin, hold the pepper over a flame or scald in boiling water. Peel and split the pods; remove the seeds and stem. Although no pretreatment is absolutely necessary, it is better to steam the peppers for about ten minutes.

Spread the peppers in a single layer on the drying trays. Sun dry for approximately two days. In controlled heat at a temperature of 150°F peppers will require approximately six to ten hours to dry. Green chili peppers are dry when they are crisp, brittle, and medium green in color.

To rehydrate, soak one cup of dried green chili peppers in two cups of water for one hour.

Peppers, Red Chili

Select red chili peppers with mature, dark red pods. The best way to dry the peppers is to wash them, string them on a thread through the stalks, and hang them root side up in an airy room. They will take several weeks to dry.

If you wish to dry red chili peppers in the oven, wash them, remove the stems, and dry them whole in the oven at 150°F. Red chili peppers are dry when the pods are shrunken, dark red, and flexible.

To rehydrate, soak a cup of dried red chili peppers in two cups of water for two hours.

Persimmons

Persimmons can be dried with the skin on or off. Harvest unbruised, firm, fully ripe persimmons that are just beginning to soften, but are not soft enough to eat. Fruit about three inches in diameter is best for drying, but smaller fruit may also be used.

Preparation and pretreatment varies. One method is to wash the persimmons in a solution of one tablespoon of household bleach to one gallon of water, puncture the skins by rolling the fruit over a stainless steel grater, and then knead the fruit to soften it slightly. The other method is to strip the skins from the persimmons with a very sharp, stainless steel knife up to and just under the calyx (the husk and stem) at the top of the fruit.

After the fruit is prepared with either method, tie the persimmons in pairs onto short pieces of string, tying the ends of the string around the calyx of the persimmons. Suspend the pairs over a rack set high in an oven heated to 110°F to 120°F. In two or three days, the persimmons will be dry, shriveled, and will no longer look or feel moist. Persimmons may scorch near the completion of drying, so reduce the temperature at the end to prevent this from happening.

Pack persimmons in an airtight container and store in a freezer, refrigerator, or cold storage area. Persimmons will keep for as long as ten months in a refrigerator. They will mold if stored at room temperature. After six or eight weeks in a refrigerator, the fruit will become whiter as natural sugars accumulate on the surface.

Plums

The best varieties of plums for drying are Abundance, Burbank, Clifford, Hunt, Santa Rosa, Beauty and Satsuma. Plums may be dried whole with their pits, halved or sliced. (Prune plums are considered in a separate entry that follows.)

The skins of small, dark plums to be dried whole will generally have to be cracked by plunging the fruit into boiling water for 15 to 30 seconds, then immersing it in cold water. Remove, drain and pat dry. As for pretreatment, halves and slices may be steam-blanched—halves for 15 minutes, slices for five minutes. Plums to be dried whole do not require steam blanching, but only the skin-cracking procedure. Or the plums may be pretreated by sulfuring, using one teaspoon of sulfur for each pound of fruit. Sulfur whole fruit for two hours, halves and slices for one hour.

Arrange the fruit in a single layer on the drying trays, pit side up for the halves. Be careful not to spill any juices that have accumulated in the pit hollows of the halves during sulfuring. For oven drying, begin with a temperature of 120°F for whole plums and 130°F for halves and slices, gradually increasing to 150°F and reducing to 140°F during the last hour when the plums are nearly dry. Do not turn the plum halves until the juices in the pit hollow have disappeared.

The average drying time for whole plums is 14 hours in controlled heat, eight hours for halves and six hours for slices. When dry, plums are pliable, leathery, and a handful will separate quickly after being squeezed.

Condition, pasteurize, package and store.

To rehydrate, soak one cup of dried plums in 1 1/2 cups of water for six to eight hours.

Plums, Prune

The best varieties of prune plums for drying are French, Imperial, Sugar, Standard, Fellemberg, Burton and Robe de Sergeant. The prunes selected should be fully ripened, firm and juicy, with flesh free from discoloration. Underripe prunes turn from purple to a dark brown and become puffy.

Prunes may be dried whole or in halves. Crack the skins of whole prunes by plunging them into boiling water for 15 to 30 seconds, then immersing them quickly in cold water to cool. Drain and towel dry. If desired, the pits may be removed from prunes to be dried whole; they should always be removed, of course, from those dried in halves.

Arrange the prunes in a single layer on the drying trays. For oven drying, begin with a temperature of 130°F, gradually increasing to 155°F, and reducing the temperature to 140°F during the last hour when the prunes are nearly dry.

When dry, prunes will be pliable and leathery. Underdried prunes will seem too soft; overdried ones will be hard and have a dull cast. The average drying time for prunes in controlled heat is 18 to 24 hours. Sun-dried prunes will take seven to eight days.

To rehydrate, soak one cup of prunes in 1 1/2 cups of water for six to eight hours.

Potatoes, White (Irish)

Potatoes are better kept in storage than dried. But if you wish to dry potatoes, select those in good condition for table use.

Peel the potatoes, rinse them in cold water and shred them for drying. Or, peel the potatoes and cut into 1/8-inch slices or shoestring strips about 3/16-inch thick. Rinse in cold water after slicing. Whether shredded or sliced, pretreat the potatoes by steaming them four to six minutes or until tender.

Spread the potatoes not more than 1/2-inch deep on drying trays and place in a 130°F oven, raising the temperature to 150°F after one hour. Reduce the temperature when drying is nearly completed to prevent scorching. Potatoes require eight to ten hours to dry and will be brittle chips when finished.

To rehydrate, soak one cup of dried potatoes in two cups of water for 90 minutes.

Potatoes, Sweet or Yams

Select firm, smooth sweet potatoes or yams for drying. Wash, peel, trim and cut them into slices 1/8 to 1/4-inch thick. Keep the slices uniform in size to facilitate uniform drying. Steam the slices until tender but not mushy. Or, steam the sweet potatoes whole with the skins on for 30 to 40 minutes until cooked through but not mushy. Then trim, peel and slice or shred.

Place the sweet potatoes on drying trays in a 130°F oven for the first hour, raise the temperature to 150°F and reduce to 130°F when drying is nearly complete to prevent scorching. Sweet potatoes will be tough to brittle when dried. Condition, pasteurize, package and store.

The procedure is the same for sun drying.

Primroses

To dry primroses, pick the flowers when in full bloom, arrange them on a drying tray with a mesh screen bottom, and dry in a warm, well-ventilated room.

Fresh-picked, primroses may also be candied. Add one part sugar to two parts water and simmer over low heat until a thick light syrup is formed. Place the flowers in the syrup and stir gently until they absorb all of the syrup. Dust the flowers with granulated sugar, allow to dry thoroughly, and store in a cool, dry place.

Pumpkin

Select deep-colored varieties of pumpkin with firm, solid flesh. Cut the pumpkin into strips about 1-inch wide. Peel off the rind and scrape off the fiber and seeds. Save the seeds for drying. Cut the peeled strips crosswise into pieces 1/8 to 1/4-inch thick and steam for six to thirteen minutes, the longer time for the thicker slices. Steaming is completed when the pieces are slightly soft, but not sticky.

Arrange the pieces in a thin layer on the drying trays. For oven drying, start at a temperature of 120°F and increase gradually to 150°F after one hour. Reduce the temperature to 130°F when the pumpkin is nearly dry to prevent scorching. Pumpkin will take 10 to 12 hours to dry in controlled heat. When dry, it will be leathery or tough to brittle and show no moisture when cut in half. Treat pumpkin in the same manner for sun drying. Condition and pasteurize both oven and sun-dried pumpkin, package and store.

To rehydrate, soak one cup of dried pumpkin in three cups of water for 90 minutes.

Pumpkin Seeds

A Halloween jack-o'-lantern will provide a plentiful supply of pumpkin seeds for drying.

Spread the cleaned seeds in a thin layer on a cookie sheet and put them in the sun for a few days to dry. Stir occasionally to prevent mold. If desired, add salt to taste. The seeds may also be dried in a 150°F oven for one to two hours or until brittle. Stir frequently to prevent scorching.

To roast pumpkin seeds, lightly coat them with oil, butter or egg white, salt them, spread them in a thin layer on a cookie sheet, and place them in a 250°F oven for 10 to 15 minutes, stirring frequently until dry.

Rosemary

During the Dark Ages, superstitious Europeans believed rosemary had the power to ward off evil spirits. Today, it is used to season meats, stews, soups, salads and a variety of other dishes.

Gather rosemary leaves at any time before the flowers appear. Dry them on a tray in a warm, well-ventilated, dry room. Crumble the leaves when dry, bottle, and store in a cool, dark, dry place.

Rosemary leaves can also be used for a tea.

Roses

In addition to their beauty, roses have a broad range of food uses, both nutritious and decorative. Where drying is concerned, the petals, the buds and the hips all must be considered.

Rose petals to be dried should be collected before the flower unfolds. Spread the petals on a drying tray with a mesh screen bottom and leave them to dry in a warm, airy room. The fresh rose petals may also be candied. Begin by soaking them for a few minutes in cold water. Remove them carefully, drain and pat dry. Add one part sugar to two parts water and simmer over low heat until a thin, light syrup is formed. Place the rose petals in the syrup until they are completely coated. Remove them one at a time and arrange them on a rack to dry. When they are almost dry, sift granulated sugar over the petals. Package and store in a dry, cool place.

Rosebuds are dried in the same manner as the petals.

The buds may also be crystallized by simmering four cups of stemmed, whole, freshly gathered rosebuds in one cup of hot water and two cups of sugar until the sugar begins to granulate. Stir gently with a wooden spoon taking care not to damage the petals of the buds. Remove and place on a rack to dry. Again, pack and store in a dry, cool place.

Rose hips are one of nature's best sources of Vitamin C. The hip is the red berry that develops after the rose flowers fade. Rose hips may be crushed for use in tea, or ground into a powder for use in baking.

Gather the hips when they are bright red. Wash them and either dry them whole or cut them open and remove the seeds before drying. Spread the hips in a single layer on the drying trays and place in the oven at 130°F, increasing the temperature to 150°F after one hour and reducing to 135°F as the drying process is nearly complete. Treat the hips in the same manner for sun drying. Condition, pasteurize, package and store.

Rutabagas

Select crisp tender rutabagas, free from woodiness. Wash them, trim off roots and tops and peel thinly. Cut the rutabagas into slices or strips about 1/8-inch thick and steam for eight to ten minutes. Rutabagas may also be peeled, quartered, steamed for 20 to 30 minutes and then shredded. In either case, the rutabagas should be tender after steaming.

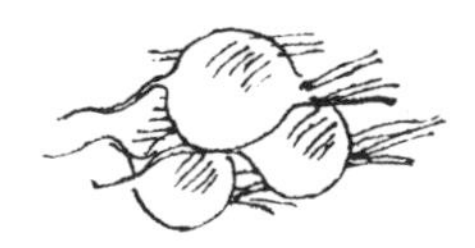

Dry the rutabagas in an oven beginning at 120°F, raising the temperature after an hour to 150°F and lowering it to 135°F when drying is nearly completed to prevent scorching. Rutabagas will need about eight to ten hours to dry in the oven and will be leathery and somewhat brittle when dried. Treat rutabagas for sun drying in the same manner. Condition if necessary, pasteurize if sun dried, package and store.

To rehydrate, soak one cup of dried rutabagas in two cups of water for 30 minutes.

Sage

This herb was prized by the ancient Romans and thought to provide long life.

To dry sage for either seasoning or tea, harvest the most succulent young leaves. Older leaves become woody and have less aromatic oils. Pick the leaves just before the plant flowers.

Dry sage leaves on a tray in a warm, well-ventilated room away from the direct sunlight. Or hang them in bunches upside down. While tray or bunch drying is preferred, sage leaves can be dried in an oven with the temperature between 100°F and 130°F. The cooler the oven, the better; warm ovens destroy the essential aromatic oils that give sage its fragrance and taste. Package in airtight containers and store in a dark, dry, cool place.

Dry sage can be used as a seasoning or can be the main ingredient in a tea. One ounce of dried sage covered with 16 ounces of boiling water will make sage tea. Flavor it with lemon and sugar or orange juice.

Sesame

Sesame seeds are found in the pods of the sesame plant, which grows in warm climates. When the pods begin to open, harvest the entire plant, tie together in bunches and suspend upside down over newspaper, cheesecloth, or a cookie sheet to catch the seeds as they fall. Package the seeds in bottles and store in a cool, dry place.

Soup Mixture

Select available fall vegetables that will yield a pleasing combination. Cut the vegetables into small pieces and prepare and dry according to the directions for each vegetable. *Never dry more than one vegetable at a time.* Combine and store the vegetables after they are dried. Satisfactory combinations may be made from cabbage, carrots, celery, corn, onions, and peas. Meat stock and rice, dry beans, split peas are usually added at the time of cooking.

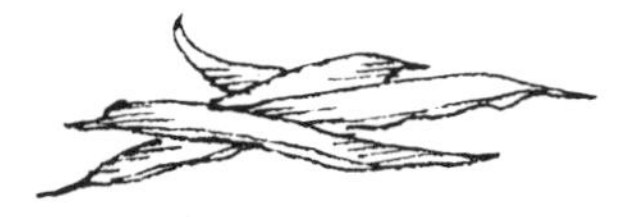

Soy Beans

Soy beans can either be dried on the vine, or picked when they are young and green for drying.

Soy beans that have been dried on the vine must be treated to kill possible insects. Spread the dry, shelled soy beans in a shallow pan to a depth of 1/2 to 3/4-inches and place in a 120°–145°F oven for three to four hours. Package the soy beans in an airtight container and store in a dry place or the soy beans may become infested with insects.

If the soy beans have been picked when young and green, shell the beans and blanch them in boiling water for seven minutes, or steam them in the pods for 10 to 15 minutes. Drain and spread the shelled beans on trays to a depth of about one inch. Begin the drying at a temperature between 115°F to 120°F, then bring the temperature up to 140°F. Stir frequently, especially at the beginning. The soy beans are dry when they are brittle and shatter when hit with a hammer.

Spinach

Select tender, crisp spinach leaves that are not fibrous. Trim off roots, wash, and dry whole leaves.

Steam spinach for four to six minutes or until wilted. Remove the mid-rib if it is prominent. Cut large leaves crosswise into several pieces to facilitate uniform drying. Spread the leaves on the drying trays so that they do not overlap.

For oven drying, begin at a temperature of 130°F and increase to 140°F after the first hour. Reduce the temperature to 130°F when drying is nearly complete to prevent scorching. Spinach requires two to five hours of controlled heat to dry and will be brittle when dried. For sun drying, prepare as described above. Condition oven dried spinach, pasteurize sun-dried. Package and store.

To rehydrate, soak one cup of spinach in one cup of water. As with other greens, spinach should be used within several weeks after drying.

Squash

Select deep-colored varieties of summer squash, winter squash or zucchini with firm, solid flesh.

Cut the squash into strips about 1-inch wide. Peel off the rind and scrape away fiber and seeds. Cut the peeled strips crosswise into pieces 1/8 to 1/4-inch thick and steam for six to thirteen minutes, the longer time for thicker slices. Steaming is completed when the pieces are slightly soft but not sticky.

Arrange the pieces in a thin layer on the drying trays. For oven drying, start at a temperature of 120°F and increase gradually to 150°F after one hour. Reduce the temperature to 130°F when the squash is nearly dry to prevent scorching. Squash will take 10 to 22 hours to dry with controlled heat and when finished will be leathery, somewhat brittle, and show no moisture when cut in half. Prepare in the same manner for sun drying. Condition, pasteurize, package, and store both oven and sun dried squash.

To rehydrate, soak one cup of dried squash in 1 3/4 cups of water for 45 minutes.

Summer Savory

The flavor of summer savory is used to enhance the taste of tomato and other vegetable juices, stews, soups, meat, fish, poultry, eggs, salads, stuffings, and cooked vegetables.

Cut the top six inches of the plant when the blooms appear. The cuttings may be tied together in bunches, suspended upside down and dried, or can be spread on trays to dry in a warm, well-ventilated room away from any direct sunlight.

When the leaves are dry, remove them from the stems, pack the leaves in an airtight container and store in a cool, dark, dry place.

Sunflower Seeds

The easiest way to dry sunflower seeds is to leave them in the flower and let nature do the drying work. You might have to help it along if you have any birds around who would like to harvest your seeds for you. If birds are a problem, wrap the flower with cheesecloth until the seeds are dry. To rush the process, shake the seeds from the dry head, spread them in a shallow pan and place in the sun for a few days.

After the sunflower seeds are dry, they can be roasted —preferably just before they are to be eaten. Spread the sunflower seeds on a cookie sheet and place them in the oven at 300°F for 15 minutes. If you add a bit of oil the color will be a fuller, deeper brown.

Tarragon

Harvest tarragon in the spring and early summer when the plant is approximately one foot high. Dry on a tray indoors in a warm, well-ventilated room away from direct sunlight. Strip the dried leaves from the stems, crush, package and store in a dry, cool place.

For tarragon vinegar, place one cup of crushed dried tarragon leaves in one cup of vinegar. Let the vinegar stand for several weeks. Strain the vinegar into a bottle, or leave the tarragon in the vinegar and strain as used. Tarragon vinegar is also made by bottling a large sprig of fresh tarragon in a pint of wine vinegar for a month or longer.

Thyme

Dried thyme leaves can be used to season chowders, soups, meats, sauces, vegetables, stuffings, salads, and as a main ingredient in an herb vinegar.

Harvest leafy stems from the plant, gather them into bunches, suspend inside a large paper bag that does not touch the herb, and suspend upside down in a warm, dry, well-ventilated room. When dry, crumble the leaves in the paper bag, bottle and store them in a dry, dark, cool place.

Tomatoes

Tomatoes are dried primarily for use in stewing. They do not rehydrate well for any other use.

Select tomatoes of good color. Steam them or dip in boiling water just long enough to loosen the skins. Chill in cold water and peel. Cut the tomatoes into sections no more than 3/4-inch wide. Cut small pear or plum tomatoes in half.

Spread the tomatoes in a single layer on the drying trays. Put in the oven at 130°F, gradually increasing the temperature to 150°F and reducing to 130°F when drying is nearly complete to prevent scorching. Tomatoes are dry when they feel leathery.

Turnips

Select crisp, tender turnips free from woodiness. Wash them, trim off roots and tops and peel thinly. Cut the turnips into slices or strips about 1/8-inch thick and steam for eight to ten minutes. Or, the turnips may be quartered, steamed for 20 to 30 minutes and shredded. With either preparation, the turnips should be tender after they are steamed.

Dry the turnips in an oven beginning at 120°F, raising the temperature to 150°F after an hour and lowering it to 130°F when drying is almost finished to prevent scorching. Turnips will take about eight to ten hours to dry in the oven and will be leathery and somewhat brittle when dried. Condition if necessary, pasteurize if sun dried, package and store.

To rehydrate, soak one cup of dried turnips in two cups of water for 30 minutes.

To dry turnip greens, select those in good condition for table use. Wash the leaves, remove the mid-rib if it is especially prominent or thick, and steam the leaves for five to seven minutes. Spread the leaves no more than 1-inch deep on a tray. Dry in an oven at a temperature between 125°F and 140°F. Stir often to prevent scorching. Dried turnip greens must be used within several weeks of drying to be satisfactory.

Violets

Violets can be dried for use in teas, or they may be candied. European violets are also crystallized but native American violets are too delicate for crystallization.

To dry violets, pick the flowers in full bloom, remove the stems, spread on a drying tray and allow to dry in a warm, well-ventilated room. When dry, package and store in a cool, dry place.

To candy violets, pick the flowers when they are still covered with dew and allow them to dry on paper towels. Beat one or two teaspoons of cold water into an egg white, causing as little foam as possible. Brush the flowers with the mixture, completely coating each petal. Dust the flower thoroughly with granulated sugar, arrange on a rack and allow to dry in a warm, well-ventilated room. Package and store in the refrigerator.

Walnuts, Black

Harvest black walnuts as soon as possible after they begin to drop from the tree in large numbers. Ripe nuts remaining on the tree can be knocked down with a long pole.

Remove the husks by crushing them with a hammer or under your heel, and picking them away with gloved hands. The gloves are important because the husks exude a brown liquid that is a strong and long-lasting dye.

Drop the black walnuts into a tub of water to help remove the remaining brown liquid, as well as to sort the good nuts from the bad. The bad nuts will float to the top, the good will sink to the bottom.

Spread the nuts in a single layer on drying trays and allow to dry in a warm, airy place for several days. After they have dried, store the nuts in sacks in a cool, dark place. If the nuts are shelled before storing, they must be kept in a refrigerator or freezer to prevent spoilage.

Walnuts (Carpathian, Persian-English)

Walnuts are mature as soon as the husk will cut free from the nut. They are usually not harvested until the nuts fall free from the husks and land on the ground. Walnuts lose their quality rapidly when they have fallen so it is important to harvest them soon afterwards to prevent mold, discoloration and decay. If the nuts are blown off the tree before the hulls crack, wait a week or two before harvesting to allow them to ripen on the ground.

To sort the good nuts from the bad, put them all in a tub of water and discard those that float. Walnuts should begin the drying process within 24 hours of harvesting. The nuts are usually dried in the shell.

Spread the nuts on drying trays a few layers deep. The optimum drying temperature for walnuts is between 95°F and 105°F. If the temperature rises above 110°F the quality of the nuts will be diminished. Air circulation is important to dry as quickly as possible. An electric heater with a fan may be the best way to dry walnuts. Drying time is approximately three to four days.

Walnuts are dry enough to store when the divider between the nut halves breaks with a snap. If the divider is rubbery, the nut is not fully dry.

Walnuts are often bleached to improve the appearance of their shells. The nuts should be thoroughly dry before bleaching. Bleach will not remove dirt that has been dried nor change the so-called red nut shell caused by premature harvesting.

Use regular household liquid laundry bleaches. Most chlorine bleaches are labeled with the percentage of chlorine content. To find the number of ounces of bleach to use in each gallon of lukewarm water, divide 140 by the percentage of chlorine in the bleach. For example, if the bleach has ten percent chlorine, divide 140 by ten. In this case, put 14 fluid ounces of bleach in each gallon of water.

Place the nuts in the solution and stir for three or four minutes. If the bleaching action is too slow, add a tablespoon of vinegar per gallon of solution.

Remove the nuts when bleaching is finished, drain and dry. Bleaching will still continue for a day or two at a reduced rate. The solution can be reused several times, depending on the amount of bleaching action needed for each batch. Only nuts that are completely closed between the shell halves should be bleached. And if appearance of the shell is not important, the nuts should not be bleached.

Store the nuts in sacks in a cool, dry place. If shelled before storing, the nuts must be kept in a refrigerator or freezer to prevent spoilage.

STORING

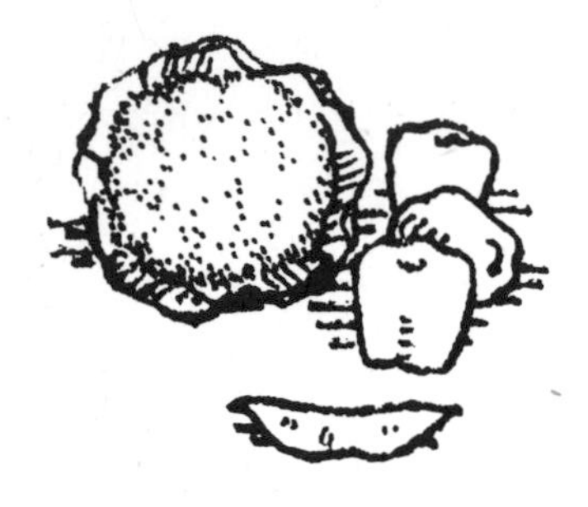

Early man generally ate whatever he could find. He left little food for the next day's meal, let alone the next month's. Immediate gratification of hunger was all that mattered. But as he began to expand his territory, moving from warmer to colder climates, he found he had to put some food aside or face starvation in winter when little or none was available.

Fortunately, the cold and dampness of primitive dwellings helped keep certain of the things he put aside in good condition. Over the centuries, he began to cultivate these—mostly root vegetables and firm-fleshed fruits—not only to eat fresh during the growing season, but also to store away for the long cold winter.

As he developed the skills to cultivate food, man also developed the techniques to store and preserve it. By the seventeenth century, the forerunners of many of the varieties of storable fruits and vegetables we know today had been pretty well established. So, too, had many of the basic principles for storing them.

As recently as the turn of the century, nearly every home had a natural area for keeping storable produce in good condition. Until that time, houses were heated principally by fireplaces and the basements or foundations were nothing more than pits dug in the soil. They were cold, dark and damp—not very pleasant for people, but ideal for storing produce. With modern technology came central heating and heated basements, a tremendous gain in human comfort but a tremendous loss where produce storage in the home was concerned.

Farm families had outdoor storage buildings or separate root cellars and most still maintain them today. The rest of us have come to rely on commercial storage facilities to keep the produce our forefathers stored themselves. As food has become a big business over the years, the reliance has become a bad bargain, with the quality of produce diminishing as its cost has increased.

Advantages and Disadvantages of Storing

Of all the methods for keeping garden produce, storage is probably the most rewarding. Without question, properly stored fruits and vegetables are the next best thing to fresh-picked. There is little change in the taste or appearance of fruits and vegetables picked at the right time and stored under the right conditions. Nor is there loss in nutritive value, except perhaps for small quantities of vitamin C in some produce.

Storage is also probably the safest method of keeping produce. Stored fruits and vegetables hold no secrets. In contrast with canned food, for example, you can tell simply by looking whether or not stored food is spoiled or free from contamination by dangerous bacteria. What's more, there is less work involved in preparing produce for dry storage than in any other method of food preservation.

Storage, too, offers the potential for keeping larger amounts of produce than other methods. Home freezers can accommodate only several cubic feet of frozen food. The cost of equipment and supplies, along with the work involved in processing, tends to limit the amounts of produce most of us choose to can. The work of pre-treating and processing, as well as the time involved, can place similar limitations on the amounts of foods we choose to dry.

While it's true that not all fruits and vegetables lend themselves well to storage, enough do (witness the individual listings in the directory that follows) to make storage a preservation method well worth considering. And the quantity of storable produce you can put away is limited only by the size of your storage facilities and the amount of good fresh produce available to you.

The matter of storage facilities brings us to one of the major drawbacks or disadvantages of the storing method. Proper facilities—those that will maintain just the right temperature, humidity, light and ventilation—are the key to successful storage. In fact, no produce can be stored for any significant length of time unless the storage conditions are exactly right. Building a facility that will provide these conditions for a wide and varied range of fruits and vegetables can be an expensive proposition. There are less expensive storage facilities, of course, but some of these are also less adequate in that they can accommodate only a limited selection of fruits and/or vegetables.

The ideal facilities, both from the standpoint of cost as well as the effect on the produce stored in them, are those that use no machinery to create and maintain the proper storage conditions. Once such facilities are built, they require care and attention if they are to do their jobs properly. In some instances, this might involve nothing more than checking a thermometer and a wet bulb/dry bulb hydrometer each day. But it also might entail trudging through drifts of snow on a bitter winter morning to adjust the ventilation in a too-cold root cellar, or digging through hard-frozen earth and straw to reach vegetables or fruits stored underground in outdoor pits.

How Storage Works

Almost all foods will spoil unless something is done to prevent spoilage. Breads mold, meats decay, fruits and vegetables rot. The objective of all food preservation is to control those elements that cause or contribute to spoilage. To store produce successfully, four factors must be controlled or regulated: temperature, humidity, air and light.

Temperature—Just as the warmth of the sun and air ripens produce in the field, continued exposure to warm temperatures will continue the ripening process to the point of spoilage. At the other end of the scale, the spoilage is much more rapid; exposure to below-freezing temperatures will quickly harm most fruits and vegetables and make them inedible. Produce placed in storage, therefore, must be kept at temperatures low enough to check the ripening process, yet not so cold as to cause frost damage.

Storage temperatures vary according to the specific fruit or vegetable to be stored. Most seem to do well in temperatures ranging between 32°F and 40°F. Some prefer temperatures in the 40°F to 50°F range. And a few fare best at temperatures above 50°F. The ideal storage temperature for each is given in the directory.

Humidity—Once fruits and vegetables are picked, they can no longer depend on the mother plant or tree for the moisture needed to sustain them. In storage, this moisture must be provided for them. As with temperature, the humidity to be maintained in a storage facility depends largely on the produce stored there. Most fruits and vegetables do best in a facility with an average humidity of 90 to 95 percent—a very moist atmosphere but a necessary one to keep them in peak condition. Without proper humidity stored produce will shrink and shrivel, lose quality and color, and eventually become unfit to eat.

Air—While fruits and vegetables need oxygen to grow and mature, fully grown produce does rather nicely with reduced amounts. In fact, the quantities of air in storage facilities must be controlled to check the process known as oxidation, a natural chemical reaction that occurs in all plants causing them to decay.

Light—Like air, light is also a necessary element for plant growth. This same element can promote continued and unwanted growth in stored produce; potato eyes will begin to grow, onions will begin to sprout, etc. Darkness in the storage facility, the easiest of all the four factors to maintain, helps prevent this kind of spoilage.

All produce is not suited to storage, and the kinds that are do not fare equally well in every kind of storage facility and condition. The more perishable varieties, thin and/or soft-skinned fruits and vegetables, understandably have the greatest sensitivity to temperatures and humidity levels outside their acceptable ranges and tend to spoil more rapidly. These, therefore, require facilities in which these conditions can be carefully controlled. Most root vegetables, on the other hand, are much less sensitive and can be kept well in facilities that may have greater temperature and humidity variations. The full range of storage facilities, along with the equipment and procedures for maintaining them, are discussed on pages 119–136.

Storage and Climate Zones

The type of storage facility you choose, indeed whether you choose one at all, depends to a great extent on where you live. For example, if you live in a warm climate where the growing season is year-round, you wouldn't need a storage facility, nor would any suffice. The ideal storage facilities, those that can accommodate a variety of produce and efficiently utilize the surrounding air and ground temperatures to provide the required storage temperatures, function best in areas with average winter temperatures in the 30°F range.

If you live in an area where the average winter temperature is in the 40°F or 50°F range, you will probably not be able to store certain kinds of produce unless you equip your storage facility with electrical cooling devices. These are expensive to install and equally expensive to operate. Either that, or you will have to limit the produce you store to those varieties that keep well at higher temperatures.

If the average winter temperature in your area is below 20°F, you will have to take special precautions to insulate your storage facility against the possibility of freezing. Even with that, you may not be able to store some varieties of produce that prefer warmer storage temperatures.

If you live in a particularly arid or dry area, you will have to find ways to provide the required amount of moisture for your storage facility, a number of which are described on page 130. On the other hand, if you're in an area that is moist and humid, you may have to take measures to keep your storage facility free of excess moisture that can cause stored produce to decay.

Preparing Produce for Storage

While commercial facilities will store produce for periods of only a few days, a home facility is most practical for fruits and vegetables that can be stored longer than two or three weeks. Given this time restriction, there are still many different kinds of fruits and vegetables that can be placed and held in storage for periods ranging from three weeks to as long as a year. Each has its own set of ideal conditions under which it will keep most successfully for the maximum length of time. These are described in detail in the individual articles in the directory section, along with the specific procedures for preparing each kind of fruit and vegetable for storage.

There are, of course, certain general rules that apply to all produce prepared for storage. Briefly, these are:

- Produce selected for storage should show no visible evidence of disease, decay or insect damage. Reject anything that is not in peak condition.
- Handle the produce as little as possible. If you choose, you may wash fruits and vegetables before storing them. This is not necessary, however, and is not recommended (except possibly for root crops) because excessive handling may cause bruises or cuts, making the produce susceptible to mold and decay in storage.
- Harvest produce for storage as late in the season as possible, but before the first killing frost. Light frost may not appear to harm some fruits and vegetables, but many do not store well if exposed to any kind of frost even though there may be no visible sign of damage.
- Fruits and vegetables to be stored should be mature or nearly mature at harvest. Late varieties, those that mature near the end of the growing season just before

the first frost, generally store better than those that mature earlier in the growing season.

- Harvest produce to be stored early in the morning, if possible, at which time fruits and vegetables have less field heat than later in the day. Before placing them in storage, let them stand for awhile in a shady area to dissipate whatever field heat they may have. Field heat in stored fruits and vegetables can quickly raise the temperature in the storage facility and cut down on the storage life of the produce kept there. If you must harvest late in the day, let the fruits and vegetables stand overnight before placing them in storage. *Never store any produce immediately after harvesting.*
- Do not store fruits and vegetables together over a long period of time. A number of fruits, apples and pears among them, give off a gas called ethylene that affects many vegetables. It will cause tomatoes to ripen, bleach cabbages from green to white, and cause carrots to taste bitter. Apples and pears also tend to absorb the flavors and odors of some of the vegetables with which they're stored. This is certainly true of onions and garlic, and over a period of time could also be the case with turnips, cabbages, broccoli and other strong-flavored vegetables.
- Do not place stored produce on concrete floors or against concrete walls because concrete promotes mildew. Set the containers on wood slats or blocks off the floor and allow several inches of air space between the containers and concrete walls.
- Carefully check and sort all stored produce in open storage facilities every few weeks, removing anything that shows even the slightest trace of spoilage.

Storage Containers

A wide variety of containers may be used for storage—wooden apple boxes, lug boxes in which tomatoes, grapes and nectarines are shipped, slatted-wood citrus or melon crates, wood and plastic barrels, lightweight tub buckets, splitwood baskets, plastic baskets, plastic bins, even plastic bags. All are acceptable, provided their inner surfaces are clean, smooth, and free of wire staples, nails and rough spots that could mar the produce placed in them and cause spoilage. The containers best suited for each specific kind of fruit and vegetable are noted in the individual articles in the directory.

Generally speaking, produce that requires some air circulation—apples, for example—should be stored in large, open-topped wooden crates. These would not do well for produce requiring high humidity. In this instance, polyethylene bags or plastic box liners are ideal. Cut several 1/4-inch holes in the sides of the bags or liners to permit some ventilation and close with a plastic or wire tie. The plastic will tend to retain the high humidity levels inside the bags or liners and the produce stored in them will not wilt or shrivel as rapidly as the same produce stored in open crates or boxes.

Containers used for storage should not be filled so deeply that the produce at the bottom might be ignored during the periodic sortings for spoilage. The bottom-most produce in containers filled to the brim may also be deprived of proper ventilation or be subjected to possible bruising from the weight of the produce packed on top. A good rule is to pack produce no more than two or three layers deep inside the containers.

Finally, storage containers and bins should not be

permanently installed in storage facilities. They should be portable and taken outside for a thorough cleaning and airing in summer when the storage facility is not in use.

Storage Facilities

Home storage facilities can be anything from a simple pit dug in the ground to an elaborate refrigerated building carefully sectioned for storing a variety of fruits and vegetables, each section equipped with individual temperature and humidity controls. Assuming the produce selected is in prime condition, successful home storage depends more on the storage facility than on any other single factor.

Storage facilities fall into two broad categories, aboveground and belowground. Aboveground units often make use of existing facilities such as garages or outside storage buildings. More affluent storage practitioners have installed walk-in restaurant refrigerators or converted saunas for storage use. It's also possible to convert a corner of the home basement into a storage facility. This is not only cheaper and easier to build but also affords more convenient access than belowground facilities. If the water table in your area is close to the surface, or if you're subjected to periodic flooding from nearby streams, aboveground storage may be your only choice. The same is true if your property is primarily rock or ledge.

Belowground facilities, such as root cellars and pits, are usually easier to control than those above ground. The ground provides natural cold and humidity, and it is a relatively simple matter to insulate against extremes of warmth and cold. But aside from pits, the larger kinds of belowground facilities are usually more expensive and more difficult to build than similar aboveground units. Ventilation and drainage can present problems. They are often harder to get to, especially in the dead of winter, and harder to keep clean.

The type of storage facility you select, therefore, will depend on several factors—where you live, the existing facilities in your home or on your property, the time and money you can spend on the storage facility, and not least of all, the kinds of produce you want to store in it.

Storing in the Home

A modern basement with its water pipes, heat vents, furnace and concrete floor can never be the kind of root cellar storage area many of our grandfathers knew. Still, it's possible to create a workable and useful storage facility in a corner of the basement in most modern homes.

Begin by making a survey of the basement and selecting an area with no heating ducts, oil or hot water pipes running through it. Heat generated by the pipes or ducts would play havoc with the temperatures in a storage area. Ideally, the area should also be as far from the furnace as possible. If it meets all the other criteria, a north or east corner is best. One final requirement: the area should have at least one and preferably two windows to provide the air flow necessary to control temperatures. Two windows are particularly desirable if the room is to be divided for separate storage of fruits and vegetables. The windows must be shaded to prevent light from entering the storage area.

The minimum area for a basement storage facility is about six feet by six feet. You may build larger if you wish, but any smaller room is not practical for efficient storage. As shown in the accompanying illustrations, the two outside walls of the house will serve as the outside walls of the storage facility. Construct the two inner

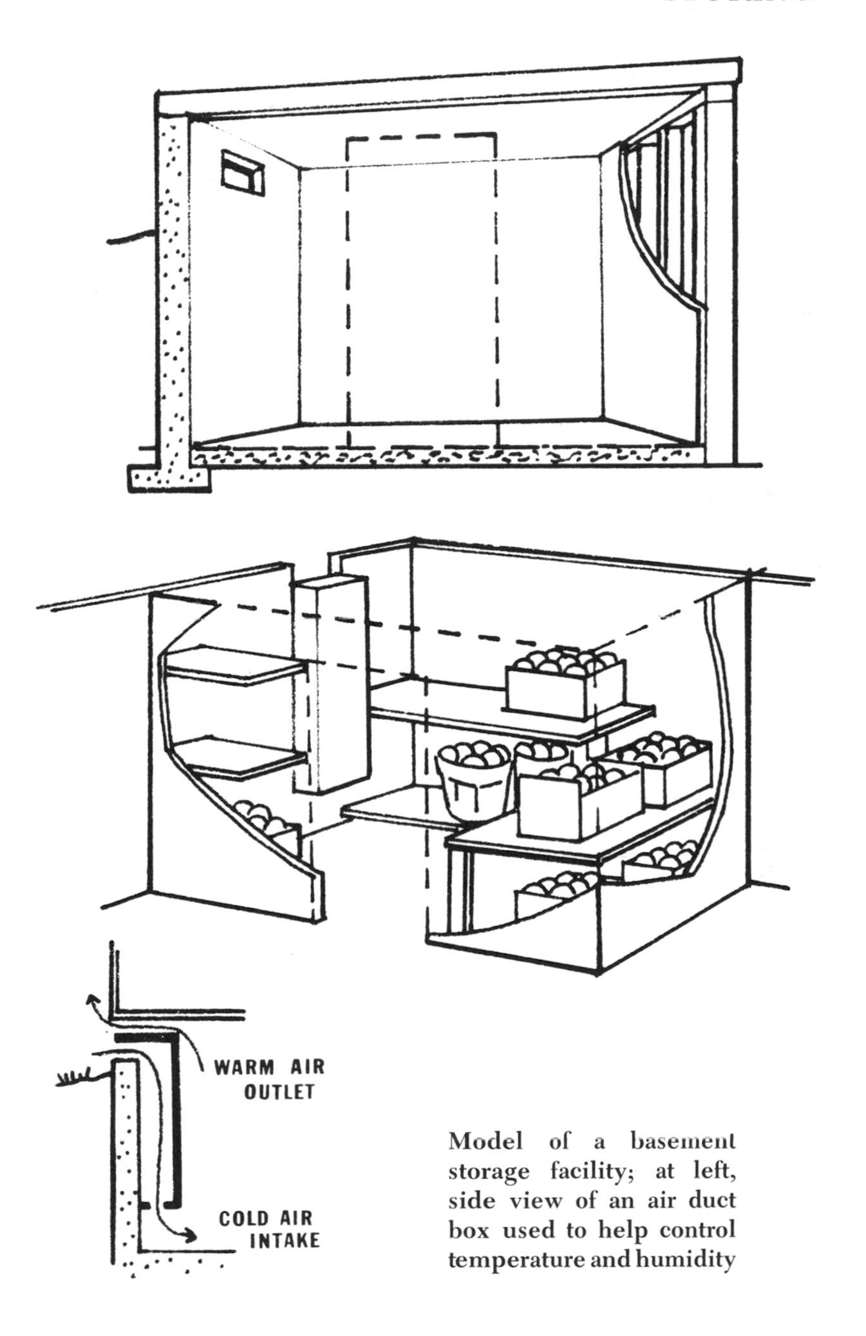

Model of a basement storage facility; at left, side view of an air duct box used to help control temperature and humidity

walls of fiberboard nailed to studding. Leave room on one of the two inside walls to hang a door.

Insulate the room with any commercial insulating material. You will usually need to insulate the two inside walls only since most outside walls have already been insulated. If yours have not been, stud the two outside walls and insulate them as well. Cover the insulation with a plastic vapor barrier and finish the inside walls with fiberboard. Add shelves of slatted wood, along with removable slatted flooring. This will not only keep storage containers off the concrete floor and help air circulation, but also allows for wetting down the floor or covering it with a layer of moist sand or sawdust to raise the relative humidity to the desired levels.

A good ventilation system that provides both air circulation and temperature control is vital to the effective operation of the storage facility. If there are two windows in the storage area, use one as an intake for cool air and the other as an exhaust for warm air. The air from the intake window should be vented to enter the room just above the floor level. The exhaust window should vent to the outside near the ceiling of the facility. When the two windows are opened at the same time, the cool air entering the room will drive out the warm air that rises naturally to the ceiling, maintaining the required cool temperatures.

If there is only one window available, construct an air-duct box to cover the lower half of the window. Carry the box down the side wall, with the opening about a foot above the floor, again as shown in the illustrations. When the window is opened, the cool air will flow through the box to enter the room at floor level, forcing the warm air nearer the ceiling out of the facility through the top of the open window.

If you don't want to go to the expense of building a new room, or if your basement will not lend itself to the type of storage facility we've discussed, you may find that you already have a small but useful storage area in your basement. Many homes have an outside entrance to the cellar. Usually this entrance has a flight of wood or concrete steps under a wide hatch door laid at a 45-degree angle to the ground. There is also an inside door opening into the basement.

The stairwell between the inner door and the hatchway can serve as a storage area for barrels or boxes of produce. You probably do not use that area for access to the outside very much in winter in any case. Simply insulate the door to the cellar with fiberglass batting or similar insulating material. Cover the floor with wet sand or sawdust to provide proper humidity. Control temperature by opening the outside hatch to let in cool air or the inside basement door to let in warm air.

While this method will not permit a large volume of storage, it is an inexpensive and simple way to store small quantities of less perishable produce.

The attic may also be used to store certain fruits and vegetables, but the temperature fluctuations in attics are generally too extreme for most produce. Even those items that can be stored—winter squash, pumpkins and onions, for example—will have to be watched carefully and will not keep as well in an attic as in other storage facilities. For attic storage, the best procedure is to build a small storage room in a northeast corner. If the room has a window, follow the same methods to ventilate the room and control temperatures as in the cellar storage room. If there is no window, the temperature may be able to be controlled somewhat by opening and closing the attic door, although this is erratic and uncertain at best.

Storing in Outbuildings

Produce can be stored in aboveground outbuildings in areas where the climate is consistently cold, but the average winter temperature is not below 32°F. Whenever temperatures in these areas drop below freezing, supplemental heat may have to be provided. Electrical heating units should be used whenever possible because the fumes from gas and/or oil units may affect the stored produce. For this reason, too, creating a storage unit in a section of a garage is not recommended. Not only is it difficult to control temperature and humidity in a building that is used for cars and machinery, but the fruits and vegetables may absorb oil and exhaust fumes that will quickly render them inedible.

Aboveground storage buildings may be constructed of masonry, lumber, or a combination of the two, but they must always be well insulated. Hollow-block cement walls, no matter how thick, afford little insulation; the channels of the blocks should be filled with some dry granular insulating material as each course is laid. Cinder blocks should be scrubbed on both sides with cement grout to make them less porous and give them some insulating value, and painted inside with aluminum paint as a moisture barrier. Lay tar paper between the ceiling and joists, again as a moisture barrier, and spread at least 12 inches of granular insulating material in the space between the ceiling and the roof for further protection.

Frame storage buildings can be built of regular 2 x 4 framing lumber and insulated with any commercial insulation. Cover the studs with building paper and use sheathing panels both inside and outside to make the building tight. Again, paint the inside with aluminum paint to serve as a moisture barrier. In all cases, be sure to install intake and exhaust vents for air circulation and temperature control.

Another serviceable storage building is one that combines belowground and aboveground features. The building is cut into the side of a hill or built with the soil banked around the three side masonry walls. The fourth wall is left exposed and has a well-insulated double door for entry. (See the accompanying drawing.)

An Outdoor Storage Building

For best results, try to face the entry door away from the prevailing winter winds. If both fruits and vegetables are to be stored, the building will have to be sectioned off into separate storage compartments. Again, air intake and exhaust vents must be installed for ventilation and temperature control.

Storing in the Garden

One of the least expensive and trouble-free ways of storing certain produce over the winter is simply to leave it in the ground. Some root crops winter very well in this way, particularly in areas that are not subjected to continuing or extreme below-freezing temperatures.

To hedge your bets a bit, take some of the root crops out of the ground and store them in a more conventional way—in a root cellar or underground pit, for example—to make sure that you have some edible produce in case the method does not work for you.

Parsnips, horseradish and salsify often improve their flavor when exposed to freezing temperatures in the garden and may be left in the ground over the winter. Celery and cabbage also lend themselves to ground storage in the garden. Specific procedures for storing in this way are described in the individual articles in the directory.

Building a Root Cellar

As the name implies, root cellars are used primarily to store root crops—carrots, turnips, beets, potatoes, etc. But root cellars can also be readily adapted to store a variety of other produce. A dirt-floored basement under a house that has not been modernized makes a perfect root cellar. But there are few of these around nowadays. Chances are, if you want a root cellar, you'll have to build one.

Cellars built entirely below ground maintain desirable temperatures longer and more uniformly than above-ground facilities. Locate the cellar as close to the house as possible and practicable. You'll not only have to visit the cellar regularly to remove produce as you need it, you'll also have to check at least once a day, probably more often, to see that the temperature and humidity are at the correct levels. If the root cellar is far from the house, you're liable not to do either as often as you should.

Once you've settled on the location, dig a hole about ten feet square and ten feet deep. These dimensions will give a facility large enough to store considerable and varied amounts of produce, yet small enough to maintain and control the temperature and humidity. As a further guide to the height of the facility, allow for a seven-foot ceiling and be sure that the top of the facility, after the roof has been installed, is at least four to six inches below the frost line—the depth to which the ground usually freezes. The local weather bureau or extension service can tell you the frost line for your particular area. For the best insulation, there should be at least two feet of earth above the roof of the cellar.

Drainage conditions will dictate the way the floor of the root cellar is finished. If the drainage is poor, cover the soil at the bottom of the pit with six to eight inches of crushed rock and lay a four-inch slab of reinforced concrete over the rock base. If the drainage is good, there is no need to lay a floor at all. In fact, a dirt floor is by far the best if conditions permit; the breathing soil helps maintain the proper moisture conditions in the cellar. Simply rake the dirt smooth and pack it down hard.

Reinforced concrete is the best material for the cellar walls. Cement or cinder blocks may be used, but the hollow channels inside should be filled with a granular insulating material just as in the aboveground facility. The exterior should be waterproofed with an asphalt sealer and the interior given a coat of aluminum paint as an additional moisture barrier. Water seepage into the cellar can quickly cause the produce to mildew.

The roof should be a reinforced concrete slab to provide enough strength to support the weight of the earth over it. Cover the top with an asphalt sealer and spread a layer of asphalt paper for additional drainage protection before filling in with the earth.

The accompanying drawing shows a model of an underground storage cellar. A hatchway door placed at a 45-degree angle to the ground and leading down a flight of wood or concrete stairs provides the simplest and best entry to the cellar. The door should be well insulated with fiberglass batting or any other good quality commercial insulating material, and faced away from prevailing winter winds if possible to guard against extreme temperature fluctuations.

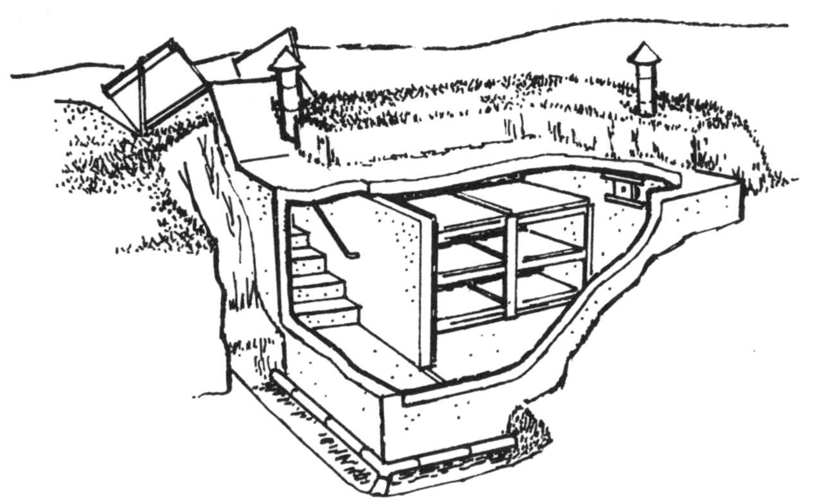

Model of an Underground Root Cellar

The ventilation system in a belowground root cellar utilizes two vent pipes about four to six inches in diameter. The inlet pipe that brings cool air into the storage area should extend from just above the normal snow line to within six inches of the cellar floor. Cover the top of the pipe with a ventilator hood, the kind that revolves and brings in air when the wind blows. It should also have some sort of damper inside so that the opening can be controlled.

The outlet pipe should extend from just above the

normal snow line to the ceiling of the storage area. Place a fixed hood over the top of the pipe—one that can be opened and closed—and install a controlling damper on the section inside the storage area. Wire screening over the outside ends of the intake and outlet pipes will keep out birds and small animals.

Controlling Conditions in Storage Facilities

The temperatures at which fruits and vegetables can be stored vary greatly. Many will keep adequately if the storage facility is maintained at temperatures between 32°F and 40°F. Some, however, prefer temperatures between 40°F and 50°F. And a few fare better in even higher storage temperatures. The specific temperature and humidity requirement for each storable fruit and vegetable is given in the individual listings in the directory.

The storage atmosphere in the facilities described thus far is maintained through ventilation that depends on outside or natural conditions and uses no electrical or mechanical energy. Temperature and, to a lesser extent, humidity are controlled by opening and closing vents, doors and windows.

During warm winter afternoons, for example, a vent can be opened to let warm air into the storage area, thereby raising the temperature if needed. Opening the same vent in the colder evening temperatures will have the effect of pushing out the warmer air and cooling the storage area. Daily adjustment of the vents is necessary to maintain proper temperature ranges in the storage facility.

In very cold areas, the storage facility must be well insulated in order to keep the produce from freezing. Much of the insulation is provided by the material with which the facility is built. But it may also be necessary to add insulating materials such as leaves or mulch, grass or straw, to keep the facility at the correct temperature.

To aid in keeping temperatures at necessary levels you will need at least two thermometers of the type that record both maximum and minimum temperatures. One should be placed at the coldest location in the storage area to assure that no part of the area is below the minimum temperature desired. The other should be placed outside the storage area to record the outdoor temperature. The fact that the air temperature outside a storage area is below freezing does not necessarily mean that the belowground temperature will be below freezing as well. But the outside thermometer can warn you of extreme drops or rises in temperature that may affect the temperature in the storage area.

As with temperature, the humidity maintained in the storage area depends largely on what is stored there. Most fruits and vegetables do best in a storage facility with an average humidity of 90% to 95%—a very moist atmosphere, but a necessary one to keep stored produce in peak condition. Without proper humidity stored fruits and vegetables will shrink and shrivel, lose quality and color, and eventually become unfit to eat. A wet bulb/dry bulb hydrometer should be placed in the storage facility to give humidity readings.

There are a number of ways to manipulate conditions in the storage area to maintain proper humidity. To raise the humidity:

1) Sprinkle the floor of the storage area with water as often as necessary;

2) Place large pans of water under the air intake vents;

3) Place wet material such as straw or odorless sawdust on the floor of the storage area and keep it moist;

4) Store produce in polyethylene bags and box liners punched with 1/4-inch holes for ventilation;

5) In an area where the humidity is consistently low

and difficult to raise to the proper level, pack fruits and vegetables in moist sand, leaves or hay (only as a last resort, however).

It is unlikely that the humidity in most storage areas will rise naturally above 90% or 95%. Still, if it should become necessary to lower the humidity, place small candles or a candle-burning lantern in the storage area for a few minutes. Watch them carefully, however, for they will also tend to raise the temperature.

Air flow is critical to maintaining temperature and humidity in storage facilities. Produce placed in storage gives off a certain amount of heat that will raise temperatures. This warm air must be exhausted from the storage area before the temperature rises above the maximum permitted for the produce stored there.

The exhaust is a relatively simple matter utilizing the system of cold air intakes and warm air outlets described previously. As cold air is vented into the facility at the floor level, it will force out the warmer air that has risen naturally toward the ceiling. As a general rule, about 60 to 80 square inches of ventilating flue is required for every 1,000 cubic feet of storage space.

Storing in Mounds, Pits and Boxes

If you have only a few crops to store or have no wish to build and maintain a large storage facility, storage in mounds, pits and boxes may be worth considering. To use this method successfully, however, the winter temperatures in your area must average around 30°F. The stored produce may spoil if temperatures are much higher and freeze if they are much lower. The method is also only useful in areas where there is little or no surface ground water and reasonably good drainage.

Besides temperature and drainage considerations, there are other limitations to these storage facilities. For one thing, they can only accommodate small amounts of produce. There is no way to control or adjust temperature and humidity conditions. And once a mound or pit is opened, all the produce must be removed for it cannot be safely reclosed.

On the positive side, mounds and pits are simple and inexpensive to build, need little maintenance, and can store certain kinds of produce very effectively. Even the fact that they store only small amounts can be overcome by having a series of separate mounds or pits, storing in each enough produce for a week or two.

To build a mound for storage, start at ground level and spread a thick layer of straw, leaves, or other bedding material. Lay a piece of hardware cloth over the bedding material and stack the fruits or vegetables in the center in a cone-shaped pile. While fruits and vegetables should not be mixed in the same mound, apples and pears may be combined in a single mound, as may various kinds of root vegetables. The hardware cloth provides protection against small animals and rodents who may find the food cache an attraction.

Aboveground Mound Storage

Once the produce has been stacked into a cone, carefully pull up the corners of the hardware cloth to cover the stack. Place another layer of bedding material over the cone and cover the entire mound with three or four inches of soil. Firm the soil by patting with the back of a shovel. Dig a shallow drainage trough around the mound so that surface water will be carried away and not seep through to harm the produce stored inside.

To ventilate the mound, let the bedding material around the produce extend through the soil at the top of the pile, as shown in the accompanying illustration. Cover the top with a board or piece of sheet metal to protect the stored crops from the rain. A stone will serve to keep the cover in place.

To build a pit, dig a shallow hole in the earth about six to eight inches deep. Follow the same basic directions for building a mound—layer bedding materials in the bottom of the pit, spread hardware cloth on top, and arrange the produce on the hardware cloth. Again, observe the same restrictions about mixing fruits and vegetables in the same pit. After the produce has been placed inside, cover it with the hardware cloth and add another layer of the bedding material. Then, cover the pit with several inches of soil for insulation.

To ventilate larger pits, place two or three boards or stakes up through the center of the pile of produce to form a flue. Cap the flue with two pieces of board nailed together at right angles to form a kind of a roof. As it gets colder, add more insulation to the outside of the pit, using the same bedding material used on the inside of the pit.

Box storage is a variation of the pit method. To begin, construct a box about six feet long, three feet wide and two feet deep. Bury the box in a pit dug in the soil so that the top is below the frost line, but as close to the surface as possible. Make the frame of the box out of 2 x 4 framing lumber. Stretch hardware cloth inside the frame and attach the cloth to the frame with nails or a staple gun. Line the box with 2-inch styrofoam insulation. Top the box with an insulated wooden lid fitted as tightly to the sides as possible.

Belowground Box Storage

Spread a layer of builders' sand or sawdust in the bottom of the box and place a layer of produce gently inside. Cover with a layer of sand or sawdust and put in another layer of produce. Continue in this manner until the box is filled. Put the top on the box, cover with straw, then follow with a layer of soil. In areas where the temperatures do not drop too far below freezing, you may be able to eliminate the top layer of soil to make removal of the stored produce an easier matter.

As you pack these facilities with produce, keep a chart of where you place certain produce inside the pits or boxes. In this way, you'll be able to find exactly what you're looking for with the least amount of bother. You might also want to draw a map locating the pits, mounds and boxes on your property. In deep snow areas it's probably a good idea to mark the locations with stakes that extend higher than the likeliest snow cover.

Storing in Barrels and Tiles

Simple barrels and large drain tiles may also serve as efficient produce storage facilities.

Barrel storage may be done above or below the ground. If you live in an area where there is surface ground water and the average winter temperature does not climb much higher than 35°F to 40°F, aboveground barrel storage may be preferable. In this instance, dig a slight indentation in the ground to secure the barrel so that it will not roll. Cover the barrel with layers of straw and dirt to a depth of about six inches, as shown in the accompanying illustration. Make sure

Aboveground Barrel Storage

that the opening of the barrel faces away from the prevailing winds, and that the opening is covered with enough insulating material so that freezing winter temperatures will not harm the produce stored inside. As you use the produce in the barrel pack the straw and dirt farther into the opening to keep a tight seal.

If you live in colder areas where below-zero temperatures may tend to freeze and ruin produce, barrel storage may still be possible if the barrel is placed under the ground. Bury the barrel on end, with the open top towards the surface of the ground, making sure that the storage portion is sufficiently below the frost line. Pack produce into the barrel with alternating layers of sawdust or sand, using straw on top of the opening as insulation.

A tile is a builder's term for a large, round drainage pipe. For produce storage it is buried in the ground in much the same way as described for belowground barrel storage—set on end, with the opening below the frost line. Three bushels of produce can be stored in a drainage tile 18 inches in diameter and 30 inches high. Lay the produce into the tile in alternating layers with sawdust or sand in between, and cover the top with straw or other insulating material.

The same restrictions against mixing fruits and vegetables apply to produce stored in barrels and tiles.

DIRECTORY

An alphabetical listing of fruits and vegetables that can be stored in home facilities, along with procedures for treating each

Apples

Late maturing apples are best for home storage. These include yellow Newton, Arkansas, York Imperial, Winesap, Baldwin, Spartan, Cortland, Monroe, Northern Spy, Rhode Island Greening, Rome Beauty, Spigold, Idared, and McIntosh.

Choose firm, mature but slightly underripe fruits picked in early morning before the sun warms them. Ripe apples will not store longer than two to three weeks.

Discard any apples with bruises or glassy spots known as "water core". Do not wash the apples; handle as little as possible and with care to avoid bruising them.

It is always sound practice to separate fruits and vegetables in storage. Apples give off a gas known as ethylene that causes carrots to taste bitter, cabbage to lose its color, and tomatoes and bananas to ripen. Apples will also pick up odors from strong smelling fruits and vegetables and should be stored accordingly.

Apples can be stored in boxes or on shelves. In the latter case, they should also be placed in perforated plastic bags or box liners to maintain the moderate humidity (85–90%) apples need to avoid shrinkage. The bags or liners should not be sealed but should be punctured with about a dozen 1/4-inch holes.

The best storage temperature for apples is as close as possible to 32°F. They will freeze at 30°F or below and will ripen and rot quickly as the temperature rises above 45°F.

Under good storage conditions, with temperature and humidity maintained, apples can be stored successfully until early spring. It is important to sort them periodically and remove any fruit that is not holding.

Apples may be stored in either outdoor or indoor facilities, keeping in mind the separation of fruits and vegetables referred to above.

Artichokes

Artichokes are almost exclusively grown in California but there may be an opportunity to acquire some for home storage.

The artichokes selected for storage should be firm and mature but not too large or they will be tough and stringy. Place the artichokes on slatted shelves or loosely pack them in boxes. Do not pack more than three layers deep.

Store the artichokes at a temperature of 40°F to 45°F with a humidity of 95%. Stored at the proper temperature and humidity, artichokes will keep for as long as six weeks. They are not suitable for outdoor storage facilities.

Artichokes, Jerusalem

This edible tuber is neither an artichoke nor does it come from Jerusalem. It is a member of the sunflower family and is sometimes called Canada potato.

Pull the Jerusalem artichokes in the morning before the sun heats them. The tops should be cut off and the artichokes left unwashed—just lightly brushed free of dirt.

They can be stored in slatted boxes in a root cellar, but not too many together in a box because they tend to generate heat. Jerusalem artichokes should be kept at a temperature of 32°F with a humidity of 95%.

Stored at the proper temperature and humidity, Jerusalem artichokes will keep until late winter. They may be stored successfully in either outdoor or indoor facilities.

Avocados

Avocados are grown in limited areas but there may be an opportunity to acquire some for home storage.

Handle the avocados very carefully and as little as possible. Select only those avocados which are firm and mature. Place the avocados in shallow boxes and, considering the delicacy of the fruit, it may be practical to wrap each one individually. They should be packed in single layers.

Store the avocados at a temperature of 40°F to 45°F with a humidity of 95%. Stored at the proper temperature and humidity, avocados will keep for as long as four weeks. They are not suitable for outdoor storage facilities.

Beans, Shell

Use beans that will be ready for harvest in the fall. Navy, kidney, lima, pinto, marrowfat, French Horticultural and soybeans are the best to store.

Either pick mature pods and put them in a warm place to dry or pull and dry the entire bean plant after most of the pods are ripe. After they are dry, pretreat the beans to eliminate possible weevils by one of the following methods:

1) Place the beans in a 0°F temperature or below for three or four days;

2) Heat the beans in a 180°F oven for 15 minutes; turn off the oven and leave them inside for one hour.

Package the beans in plastic bags or glass jars and store at a temperature between 32°F to 50°F. The humidity should be a dry 70%.

Beans stored under the proper conditions of temperature and humidity will keep for one year or longer. Indoor facilities are best for shell beans and they can even be stored in an unheated attic.

Beets

Beets can be sown continuously throughout the growing season, so it's possible to arrange planting that will reach maturity at just the right time for storage in late fall. The best winter storage beets are Detroit Dark Red, which takes approximately 60 days to mature, and Lutz Green Leaf, which matures in approximately 80 days.

The beets will store best if they are mature but not woody. Dig them when the soil is dry and brush the soil off very gently. You may wash them if you wish but there is a risk of bruising them by extra handling. Whichever you choose, lay them in a cool place to dry and lose the heat of the soil. Cut off the tops to within 1/2 inch of the vegetables and leave the tails.

The ideal method of storage to retain maximum crispness is to bed the beets in layers of moist sand, peat or sphagnum moss. They can also be stored loose in plastic bags or plastic box liners which have been perforated with about four 1/4-inch holes.

Beets should be stored at temperatures between 32°F and 40°F. Storage at 45°F or above can cause them to sprout new tops and become woody. They require a humidity of about 95% to prevent shriveling.

Beets stored under the proper conditions of temperature and humidity will keep as long as five months. They should be sorted periodically to remove any that have not held. Beets may be stored in either outdoor or indoor facilities.

Broccoli

Broccoli should be harvested before the first frost. Remove the large heads in the center and the smaller sprouts along the sides. Do not use any broccoli for storage that has begun to flower.

Place the broccoli in boxes with perforated plastic liners and alternate stalks and heads. They may be layered, one on another, to a depth of no more than four layers.

Broccoli should be stored at a temperature of 32°F and with a humidity of 95%. Stored at the proper temperature and humidity broccoli can keep as long as one month. Outdoor storage facilities are not recommended.

Brussels Sprouts

One of the best varieties of Brussels sprouts for storage is the late-maturing and frost-resistant Jade Cross Hybrid. It can be planted all through the summer and requires 80 days to ripen, so plantings can be arranged to mature just at the right time for storage in late fall.

Pull up the entire mature plant and remove any bruised sprouts or leaves. Pack the plants upright and close together in a plastic bin or box and cover the roots with moist soil.

Although Brussels sprouts are less pungent than cabbage, they should be stored away from any fruits or vegetables that might pick up their odor. The temperature should be maintained near 32°F and the humidity near 95%.

Stored under proper temperature and humidity conditions the Brussels sprouts will keep for over five weeks. Brussels sprouts keep equally well in indoor or outdoor storage facilities. They may be stored outdoors in the same manner as cabbages.

Cabbage

Good winter storage cabbage varieties are Penn State Ballhead and Late Flat Dutch.

Cabbage can be left in the garden until late fall since light frost does not harm it. There are a number of choices for cabbage storage.

For storage in a root cellar or dry shed, pick the mature and firm cabbages, roughly trim the heads and cut off roots. The cabbages can be placed in boxes with a covering of moist soil or sand, or they can be stored in slat boxes or plastic bags with a few holes punched in them for ventilation. It is not a good idea to store cabbages in a basement root cellar as they tend to perfume the entire house with their strong odor. Ideally, they should be stored in root cellars removed from the home.

If you intend to use mound storage, leave the cabbage heads untrimmed and the roots intact. Place the plants side by side, root-side up, along the area to be used. Cover the cabbages completely with soil, mounding it well, and place a plastic cover over the top. Drainage is important and a ditch should be cut around the mound to facilitate run-off for the water.

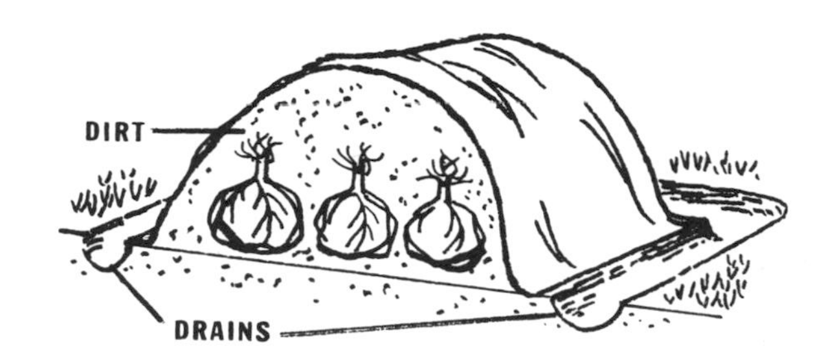

Cabbages Stored in a Mound

Still another method of storage is the plank and pole method. For this method the cabbages are pulled with their roots intact and heads left untrimmed. They are placed side by side, roots down, in a shallow trench. A recommended length for the trench is eight to ten feet and enough trenches should be dug next to one another to accommodate the cabbage harvest. A frame of poles should be erected around the bed to a height of two feet. Bank soil all around the poles and lay planks over the top, making a platform over the cabbages. Cover this platform with hay or straw. The cabbages should be cut off with a knife and the roots left in the ground. The roots will sprout in the spring and give a good supply of greens.

Plank and Pole Storage for Cabbages

Do not store cabbages near apples; the ethylene gas given off by the apples will cause cabbages to lose their color. Cabbages should be kept at a temperature of about 32°F to 40°F with a humidity of 95%. Stored under these conditions they will keep well until early spring.

Carrots

Carrots can be planted throughout the growing season and most varieties require between 70 to 75 days to mature. Although most carrots store well, some particularly good varieties are Gold Pak, Spartan Bonus and Royal Chantenay.

Pull the firm, mature and unblemished carrots in the morning before the sun heats them, rinse or brush them clean and cut off the tops. Do not store carrots next to apples because the ethylene gas given off by the apples will cause the carrots to taste bitter.

Carrots can be placed in boxes in alternate layers with moist sand. Another method is to place a base layer of newspaper in a box, then a layer of peat moss, a layer of carrots, another layer of peat moss, another layer of carrots and so on with alternate layers, ending with peat moss.

It is a good idea not to store too many root vegetables together in one box; they tend to generate heat in large quantities and that will harm the quality. It is better to prepare a number of smaller boxes than one large one.

Carrots store well in a temperature between 32°F to 40°F with a humidity of 95%. Carrots stored with these requirements in mind will keep as long as five months. Remember to check periodically for those that may not hold well. Carrots may be stored in either indoor or outdoor facilities.

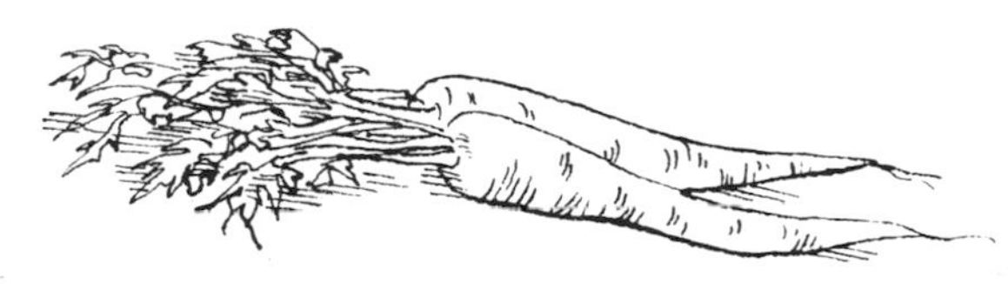

Cauliflower

Cauliflower should be planted so that it will mature in the fall since the heads do not thrive in the heat. The cauliflower requires an average of 60 days to mature and some of the best storage varieties are Early Snowball, Snow King Hybrid and Snow Crown Hybrid.

Choose mature, solid heads and brush clean. Cut off the roots but leave a good many protective leaves around the heads. Cauliflower is best stored in boxes or baskets with moist sand covering the heads.

The temperature should be as close to 32°F as possible but can fluctuate to as high as 40°F. The humidity should be at least 95% since cauliflower prefers a moist atmosphere.

Stored at the proper temperature and humidity cauliflower will keep as long as eight weeks. The vegetable should only be root-cellared and never stored in pits, barrels or other outdoor facilities.

Celeriac

Celeriac is an excellent winter storage root vegetable and can be used in soups, salads, etc.

Pull the celeriac in the morning before the sun heats the ground. The tops should be cut off and the celeriac left unwashed. The roots can be stored in slatted boxes in a root cellar but not too many together in a box as the celeriac generates heat.

Celeriac should be kept at a temperature of 32°F to 40°F with a humidity of 95%. Stored at the proper temperature and humidity, celeriac should keep until late winter. The roots may be stored successfully in either outdoor or indoor facilities.

Celery

Most varieties of celery will winter well. They should be fully mature and well-leafed on top. Pull during early morning hours before the sun heats the plants.

Celery keeps well in the garden for a month or more if protected by soil banked to the tops of the plants before the weather freezes. As the weather gets colder the entire mound should be covered with straw, hay, peat, etc. Storing by this method makes it difficult to dig the celery, however, because the frozen ground becomes very hard.

Another method is to dig a trench about ten to twelve inches wide and 24 inches deep. The entire celery plant, roots and all, should be set into the trench in an upright position with loose soil packed around the roots. The plants should be watered and the tops allowed to dry out. Set a 12-inch board on edge on one side of the trench and bank with soil. Rest a board or some poles from the ground to the top of the 12-inch board, forming a sloping roof over the celery. Place a light covering of straw or other material over this and add more covering as the weather gets colder. Unused deep cold frames may be used in the same manner as the preceding method, substituting boards for the glass in the cold frame.

For storing in a root cellar the celery should be pulled, roots and all, with some soil still clinging. Place the plants upright and close together in a box with moist sand or soil packed around them. Do not store celery near turnips or cabbage because the celery will absorb their strong odors.

Celery should be stored at a temperature as near 32°F as possible and with a humidity of 95%. Stored under these conditions the celery will keep for two to three months. Celery keeps equally well in indoor or outdoor facilities.

Chinese Cabbage

Chinese cabbage is sometimes called celery cabbage. It is a member of the cabbage, mustard, turnip, etc. family but has very little of the cabbage flavor. It is most often used in similar ways as lettuce but is a bit firmer. The outer coarse leaves can be cooked as greens.

Any of the methods used to store celery are suitable for Chinese cabbage.

Cranberries

Cranberries should be harvested in late fall. They should be washed and sorted and only firm, plump berries should be stored.

Place the berries in a box or in plastic bags. If stored

in boxes, the berries should be lightly covered. Cranberries store best at a temperature between 36°F and 40°F with a humidity between 85% and 90%.

Stored at the proper temperature and humidity, cranberries should keep for as long as three months. They are not recommended for outdoor storage facilities.

Cucumbers

Cucumbers take an average of 65 days to mature and the crops should be harvested as late as possible in fall. Some good varieties to store are Marketer, Marketmore 70 and Straight Eight.

The cucumbers should be clean and stemless. Pick them early, before the sun heats them. Pack the cucumbers closely in boxes or crates. They may be lightly coated, if desired, with paraffin before packing to minimize shrinkage.

Cucumbers should be stored at a temperature of 45°F to 50°F and with a humidity of 95%. Stored at the proper temperature and humidity, cucumbers can be kept as long as six weeks. Cucumbers are not suited for storage in outdoor facilities.

Eggplant

Good varieties of eggplant for winter storage are Black Beauty and Jersey King Hybrid.

Eggplants are mature when the skin has a high gloss. Pick the mature eggplants just before the first frost and leave a two to three-inch stem on the end. Place the eggplant in slatted boxes or on slatted shelves. Handle very carefully as they bruise easily.

Eggplant should be stored at a temperature of 45°F to 50°F with a humidity of 90%. Stored at the proper temperature and humidity, eggplant will keep as long as three weeks. They are not recommended for outdoor storage facilities.

Endive, Belgian or French

French or Belgian endive is also known as witloof chicory. The plants should be dug in late fall with some soil clinging to the roots. Tie the leaves together to help blanch them or keep them white.

Set the plants close together in a box and pack soil around the roots. The soil should be kept moist. Endive or witloof chicory should be kept at a temperature of 32°F and a humidity of 95%.

Stored at the proper temperature and humidity the plants should keep for three months. They can be successfully stored in outdoor or indoor facilities, utilizing the methods suggested for celery.

Garlic

Most varieties of garlic will store well. Pull up the garlic late in the fall when the tops have fallen over. Do not wash the bulbs but put them in a shady place for a few days to dry. Cut off the tops.

Store the garlic bulbs in tightly closed plastic bags so they won't affect other produce with their strong odor. Garlic keeps best at a temperature of 32°F with a humidity of 65% to 70%.

Stored at the proper temperature and humidity garlic will keep for up to eight months. Garlic is not suitable for storage in outdoor facilities.

Ginger

Ginger is grown only in warm climates but you may acquire some to bring back to more northern areas for storage or purchase some in markets.

The roots should be dug and set aside in an airy, shady place to dry for two or three days. Use only good-sized succulent roots. Brush the roots clean of any soil and store them at a temperature of 55°F with a humidity of 95%.

Stored at the proper temperature and humidity, ginger will keep for one month or more. Ginger is not suitable for outdoor storage facilities.

Grapefruit

Grapefruit should be handled as little as possible before storage, because any bruises will hasten decay. Choose firm, unblemished fruit for storage.

Grapefruit may be stored either unwrapped or wrapped individually. Wrapping will prevent wrinkling if the air becomes too dry and will also isolate decay, if it occurs. Place them in a slatted fruit crate or box.

Grapefruit should be stored at a rather high temperature of 60°F with a humidity of 95%. Stored at the proper temperature and humidity grapefruit will keep for six to eight weeks. They are not suitable for storage in outdoor facilities.

Grapes

The best variety of grapes for storage is Catawba, but most other varieties can be stored with success.

Grapes should be picked in the early morning before the sun heats them. Select firm, fully mature bunches and wash them carefully. Leave the grapes on their stems in bunches. Place the bunches in closed, but not airtight, containers so they will not absorb other odors.

Grapes should be stored at a temperature of 30°F to 35°F with a humidity of 90%. Stored at the proper temperature and humidity the grapes should keep as long as two months. Grapes are not suitable for storage in outdoor facilities.

Horseradish

One of the best varieties of horseradish roots for winter storage is Maliner Kren. Sets of the horseradish are planted in spring and are ready for use or storage by late fall or early winter.

Horseradish roots winter over in the ground quite well. They can be dug as needed, but the digging can be very difficult in some areas. If you prefer to pull and store them, choose firm mature horseradish roots. Cut off the tops and brush clean. Put them in boxes or perforated plastic bags.

Store the horseradish roots at a temperature of 32°F to 40°F with a humidity of 95%. Horseradish roots stored at the proper temperature and humidity will keep until early spring. Indoor or outdoor storage facilities are equally suitable.

Kohlrabi

Two very good winter storage varieties of kohlrabi are Early White Vienna and Early Purple Vienna. The edible part of a kohlrabi grows on the stem just above the ground and is about the size of a small turnip or apple. Pick the vegetables early in the day before the sun heats them and remove tops and roots. Select firm, mature kohlrabi.

Pack the kohlrabi in moist sand in boxes or plastic bins. Store at a temperature of 32°F to 40°F with a humidity of 95%. Stored at the proper temperature and humidity kohlrabi should keep for three months or more. Kohlrabi can be stored in outdoor or indoor facilities with equal success.

Leeks

The best variety of leek for winter storage is Broad London. Select mature vegetables but not overlarge or they will be tough. Pull the entire plant, roots and all, as late as possible in the fall.

Stand the plants upright and close together in box. Pack soil around the plants covering the roots and white part.

The temperature for storing the leeks should be maintained between 32°F and 40°F with a humidity of 95%. Leeks stored at the proper temperature and humidity will keep three months. Indoor or outdoor storage facilities are equally successful for leeks.

Lemons

Lemons, like all citrus fruit, should be handled as little as possible before storage because any bruises will hasten decay. Choose firm, unblemished fruit for storage.

Lemons may be stored unwrapped or individually wrapped. Wrapping will prevent wrinkling if the air becomes too dry and will also isolate decay, if it occurs. Place them in a slatted fruit crate or box.

Lemons should be stored at a rather high temperature of 60°F with a humidity of 95%. Stored at the proper temperature and humidity lemons will keep for six to eight weeks. They are not suitable for storage in outdoor facilities.

Limes

Limes, like all citrus fruit, should be handled as little as possible before storage because any bruises will hasten decay. Choose firm, unblemished fruit for storage.

Limes may be stored unwrapped or individually wrapped. Wrapping will prevent wrinkling if the air becomes too dry and will also isolate decay, if it occurs. Place the limes in a slatted fruit crate or box.

Limes should be stored at a temperature of 45°F to 50°F with a humidity of 95%. Stored at the proper temperature and humidity limes will keep for six to eight weeks. They are not suitable for outdoor facilities.

Mangoes

Although mangoes are grown in limited areas, there may be an opportunity to acquire some for home storage.

Handle the mangoes very carefully and as little as possible. Select only those mangoes which are firm and mature. Place the mangoes in shallow boxes and, considering the delicacy of the fruit, it may be practical to wrap each one individually. They should be packed in single layers.

Store the mangoes at a temperature of 40°F to 45°F with a humidity of 95%. Stored under these conditions the mangoes will keep for as long as four weeks. They are not suitable for outdoor storage facilities.

Melons

Good varieties of melon for storage are the casaba, Crenshaw, honeydew and Persian. These are slow-growing melons, taking from 90 to 120 days to mature and this must be considered when planting them for storage.

Select firm mature melons before a frost and cut off the vines. Place in slatted boxes or on slatted shelves and store in a dark area with the temperature at 40°F to 50°F and the humidity at 95%. Stored at the proper temperature and humidity the melons can keep as long as six weeks. Outdoor storage facilities are not recommended.

Onions

Good varieties of onions for winter storage are Yellow Globe, Ebenezer, Red Wethersfield, White Portugal and Southport White Globe. Onions grown from sets do not winter well.

Onion tops will begin to fall over when mature. Push over the rest of the tops and let them rest for three or four days. Pull the onion plants and cut off the tops about one inch above the onion, unless you would like to braid the onion tops together. In either case, place the onions in a shady, airy spot for the necks or tops to dry out completely.

If there are any onions that have unusually thick necks or do not dry well, remove them and use immediately because they will not keep. The onions can be stored in slatted boxes, crates or mesh bags. If braided, they can be hung from a hook or nail.

Onions should be stored at a temperature as close as possible to 32°F with a humidity of 65% to 70% to inhibit rot and sprouting. Onions need active air circulation. There are sprout inhibiting (maleic hydrazide) sprays on the market used by commercial growers but they are *not* recommended for home use.

Onions stored at the proper temperature and humidity

will keep until early spring. Indoor or outdoor storage facilities are equally successful for onions.

Oranges

Oranges, like grapefruit, should be handled as little as possible before storage because any bruises will hasten decay. Choose firm, unblemished fruit for storage.

Oranges may be stored either unwrapped or wrapped individually. Wrapping will prevent wrinkling if the air becomes too dry and will also isolate decay, if it occurs. Place them in a slatted fruit crate or box.

Oranges should be stored at a rather high temperature of 60°F with a humidity of 95%. Stored at the proper temperature and humidity oranges will keep for six to eight weeks. They are not suitable for storage in outdoor facilities.

Parsnips

The best variety of parsnip for winter storage is Hollow Crown. The parsnip is a vegetable that can be left in the ground, right where it grew, all winter long. A frost will add sweetness to it.

If, however, parsnips are to be stored, they should be dug after a first frost. Store the parsnips in boxes or crates and pack dry sawdust, moist sand, peat or sphagnum moss around them. They may also be stored in perforated plastic bags.

Store the parsnips at a temperature of 32°F with a humidity of 95%. Parsnips will keep for as long as six months when stored at the proper temperature and humidity. Indoor or outdoor facilities will be equally successful.

Pears

Kieffer, Duchess, Bosc, Comice, Anjou and Winter Nelis are harvested very close to freezing weather and are the best varieties for winter storage.

Pears should be picked when they are mature but firm. They should not be ripe enough to eat or they will keep no more than two to three weeks no matter what the temperature. Pick early in the morning before the sun heats them. Discard any pears with bruises. Don't wash the pears or handle them any more than is necessary to avoid bruising them.

Pears, like apples, give off a gas known as ethylene that causes carrots to taste bitter, cabbage to lose its color, and tomatoes and bananas to ripen. Pears will also pick up odors from strong smelling fruits and vegetables. Separate from other storage produce.

Pears can be stored in boxes or on shelves but should also be encased in perforated plastic bags or box liners to maintain the moderate humidity of 85% to 90% pears need to avoid shrinkage. The bags or liners should not be sealed but should be punctured with about a dozen small holes.

The best storage temperature for pears is as close as possible to 32°F. They will freeze at 30°F or below and will ripen and rot quickly as the temperature rises above 45°F. Under good storage conditions, with temperature and humidity maintained, pears can be stored successfully until early spring. It is important to sort them periodically and remove any fruit that is not holding.

Pears may be stored in indoor or outdoor facilities equally well, but remember to store fruits separately from vegetables.

Peppers, Sweet

The best varieties of sweet peppers for winter storage are King of the North, Yolo Wonder, Worldbeater, Merrimack Wonder, California Wonder, Bell Boy Hybrid and Fordhook. The peppers should be mature and firm with no blemishes. They are best stored while green and ripened to red, if desired, after being taken out of storage.

Pack the peppers loosely and carefully in slatted boxes or perforated plastic bags. Peppers should be stored at a temperature of 45°F to 50°F with a humidity of 95%. Stored at the proper temperature and humidity the peppers will keep up to two months. Sort periodically to remove any peppers not holding.

Because of the higher than normal storage temperature needed, peppers should not be stored in outdoor facilities.

Pineapples

Pineapples are grown in limited areas but there may be an opportunity to acquire some for home storage.

The pineapples selected for storage should be firm and mature with little or no yellow color to their skins at this point. Place the pineapples on slatted shelves or loosely pack them in boxes. Do not pack more than three layers deep.

Store the pineapples in the dark at a temperature of

40°F to 45°F with a humidity of 95%. Stored at the proper temperature and humidity, pineapples will keep for as long as six weeks. They are not suitable for outdoor storage facilities.

Plums

Some varieties of plums that will store well are Damson, Italian Prune, Pond (Hungarian Prune) and Golden Drop. Plums for storage should be mature but firm and free of any blemishes. Sort carefully but handle them as little as possible to avoid bruises.

Pack the plums loosely in slatted boxes or in perforated plastic bags. It is always sound practice to store fruits and vegetables separately because fruits are likely to pick up strong odors. Store the plums at a temperature of 45°F to 50°F with a humidity of 90%.

Plums will keep as long as six weeks when stored at the proper temperature and humidity. Outdoor storage facilities are not recommended.

Popcorn

The best winter storage varieties of popcorn are Japanese Hulless, Hybrid South American Mushroom and Creme-Puff Hybrid. Let the ears of popcorn stay on the stalks until the kernels are well dried.

Pull the shucks upward, gather in small bunches and hang the popcorn in the storage area if desired. The popcorn can also be shucked entirely and left on the ears till used. A third method is to remove the kernels from the ears and place them in a glass or plastic container.

Popcorn can be stored at a temperature of 32°F to 40°F with a humidity of 65% to 70%. Stored at the proper temperature and humidity popcorn will keep indefinitely.

Controlled indoor or outdoor facilities are best to store popcorn. When stored in the house it often becomes too dry to pop; when stored in the basement it may become too damp.

Potatoes

For winter storage of potatoes use late-maturing varieties such as Katahdin, Kennebec, Chippewa, Sebago, White Rose, Russet Burbank, Russet Rural and Norgold Russet. Use early-maturing varieties such as Cobbler, Anoka, Bliss and Triumph for storage only if you live in an area with a short growing season.

Potatoes are ready to harvest when the tops start drying out. Dig the potatoes very carefully so they are not damaged. Cut the potatoes free from the plants and carefully place them in slatted boxes. It is better to do this in the garden and thereby handle the potatoes less. Any bruised, soft or blemished potatoes should not be stored.

Store the potatoes at a temperature of 40°F with a humidity of 90% and be sure that they are not exposed to any light. If the temperature drops too much below 40°F the starch will turn to sugar and the potatoes will be too sweet. If this happens, they can be reclaimed by keeping at room temperature for two weeks prior to cooking. Do not put them back in storage.

Sort the potatoes periodically and *discard* any that have turned green. Remove any that are soft or sprouting and either use immediately or discard. There are sprout inhibiting (maleic hydrazide) sprays on the market used by commercial growers but they are *not* recommended for home use.

Late potatoes stored at the proper temperature and humidity will keep until early spring. Early potatoes stored properly will keep only three to four weeks. Potatoes can be stored successfully in either indoor or outdoor facilities.

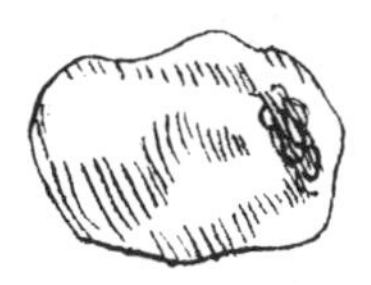

Potatoes, Sweet or Yams

Some good varieties of sweet potatoes for winter storage are Big-Stem Jersey, Yellow Jersey, Red Jersey, Puerto Rico and Nancy Hall.

Sweet potatoes and yams take an entire growing season to mature and there is no easy way to tell when they're ready. The best way is to dig up a test tuber and judge by its size. When you consider they are mature, dig them very carefully and put them into slatted boxes right in the garden. Sweet potatoes and yams are very fragile, so handle them as little as possible.

If a frost should touch the vines before you can dig the sweet potatoes or yams, chop the vines off immediately and dig the potatoes at your leisure within the next week.

Both sweet potatoes and yams must be "cured" before storage. Cure them at a temperature of 80°F to 85°F for two weeks. Keep them in the same slatted boxes as they were put into in the garden and do not handle the potatoes.

After curing, the sweet potatoes or yams should be stored at a temperatue of 50°F to 55°F with a humidity of 80% to 85%. Stored at the proper temperature and humidity the sweet potatoes or yams should keep for up to six months.

Be sure to watch for and carefully remove any sweet potatoes or yams that do not hold. Neither sweet potatoes nor yams should be stored in outdoor facilities.

Pumpkins

Small Sugar, Kentucky Field, Big Tom and Winter Luxury are very good varieties of pumpkins for winter storage. Cut the pumpkins from the vines before a frost, leaving about an inch of stem. Do not allow them to grow too large if you would like to make tasty pies. Brush off any dirt and check for blemishes or soft spots.

Place the pumpkins off the floor of the storage area. A low shelf or boards across blocks would do well. Do not let the pumpkins touch one another; this encourages rot at the point of contact. Maintain the pumpkins at a temperature of 50°F to 55°F with a humidity of 70% to 75%.

Stored at the proper temperature and humidity the pumpkins will keep for as long as five months. Outdoor storage facilities are not recommended.

Radishes, Winter

Radishes for storage should be the winter varieties such as Round Black Spanish, White Chinese or Celestial and China Rose.

Pull the entire mature radish plant and brush free of dirt. Check for blemishes. Place the radishes in slatted boxes or perforated plastic bags. Do not handle the radishes any more than necessary. If humidity is difficult to control it might be better to put the radishes in boxes and pack them with moist sand, otherwise they will dry out and wilt too fast.

Maintain the storage temperature at 32°F and the humidity at 95%. Stored at the proper temperature and humidity, radishes will keep for as long as four months. They can be stored in either indoor or outdoor facilities.

Rutabagas

Improved Purple-Top, Improved Long Island, American Purple-Top and Laurentian are good varieties of rutabagas for winter storage. The rutabaga is sometimes called a winter or yellow turnip.

Rutabagas should be harvested after a frost but before the ground freezes. Brush free of dirt, cut off the tops and select only the unblemished rutabagas to store.

Rutabagas may be dipped in melted paraffin and placed in boxes or perforated plastic bags for storage. They may also be placed in boxes and packed with moist sand. Both methods will help maintain the humidity and keep the rutabagas from shriveling.

Do not store rutabagas with other fruits and vegetables because of their strong odor. Store at a temperature of 32°F with a humidity of 95%.

Rutabagas will keep for as long as four months when stored at the proper temperature and humidity. Indoor or outdoor storage facilities work equally well.

Salsify

The best variety of salsify for winter storage is Sandwich Island. Salsify is also frequently known as oyster plant. Salsify may be left in the ground over the winter. Covered with some hay or straw, it will last until spring.

If you prefer to store the salsify, dig the plants after the first frost. Brush clean, cut off the tops and inspect for blemishes. Place the salsify in slatted boxes or perforated plastic bags in the storage area. Another method is to pack them in boxes and surround the salsify with moist sand to prevent shriveling.

The temperature for storing salsify should be maintained at 32°F and the humidity at 90% to 95%. Stored at the proper temperature and humidity salsify will keep as long as five months. Indoor or outdoor storage facilities are equally successful.

Shallots

Shallots are a milder and more delicately flavored member of the onion family. Dig up the plants when the tops wither and fall over. The bulb will separate into smaller sections called cloves. Discard the tops.

Let the shallots dry in a cool, airy place for a few days. The shallots can then be stored in slat boxes, crates or mesh bags.

Shallots should be stored at a temperature of 32°F with a humidity of 65% to 70%. Air should circulate freely around the shallots and the storage area should be dark to prevent sprouting. Stored at the proper temperature and humidity shallots will keep for as long as seven months.

Squash, Winter

Boston Marrow, Vegetable Marrow, Turban, Acorn, Butternut, Buttercup and Blue Hubbard are all winter varieties of squash that store well.

Remove the squash from the vines as they mature, leaving a two to three-inch stem. Size will be the guide to maturity. Cure the squash at a temperature between 80°F to 85°F for two weeks by leaving them in a warm room. This will help seal the stem wound and ripen any immature squash. Acorn varieties do not need to be cured.

To store, the squash should be placed side by side, not touching, in a fairly warm area. A good place is near the furnace, on slatted boards above the floor.

The ideal temperature for squash is 50°F to 55°F with a humidity of 50% to 60%. If the temperature falls below 50°F for any length of time the squash will be ruined. Stored at the proper temperature and humidity squash will keep for up to six months. Outdoor storage facilities are not recommended.

Tomatoes

Most late maturing varieties of tomatoes can be winter stored, such as Stokerdale, Marglobe, Rutgers, Manalucie, Long Red and Ramapo Hybrid. The only condition is that the vines should not be nearly spent; tomatoes from these vines will rot more quickly than those from prime production plants.

One method of storing tomatoes is to pull up the entire plant just before a first frost. Hang the plant upside down in a place where the temperature will not fall below 50°F to 55°F.

Another method is to pull the tomatoes from the vines, again just before the first frost, and sort them according to size and maturity. Do not wash or brush.

At this point a decision can be made whether to wrap them or not. Wrapping in paper, foil or plastic wrap can reduce shriveling and moisture loss but it can also hasten decay and be a bother to check periodically.

Having made the decision about wrapping, place the sorted tomatoes in shallow trays, taking care not to crowd them, and cover lightly with sheets of newspaper.

A temperature of 65°F to 70°F will ripen green tomatoes in about two weeks; a temperature of 55°F will ripen them in a month or more. Tomatoes need a humidity of 90%. Sort frequently for ripeness and/or decay.

Do not store near apples because the ethylene gas given off by the apples will ripen the tomatoes too quickly. Stored at the proper temperature and humidity tomatoes can be kept for as long as six weeks, including ripening time and storage time of ripened fruit. Outdoor storage facilities are not recommended.

Turnips

Good varieties of turnips for winter storage include Purple-Top White Globe and Tokyo Cross. Turnips can withstand a hard frost but should not be subjected to alternate thawing and freezing.

Pull turnips and cut off tops, leaving one-half inch of top attached. Brush off any dirt and discard blemished vegetables. Place the turnips in slatted boxes or in a box packed with moist sand. Turnips, like all root vegetables, dry out and wilt very rapidly if not kept in high humidity. Do not store turnips near other fruits and vegetables because of their strong odor.

Turnips should be stored at a temperature of 32°F with a humidity of 90% to 95%. Stored at the proper temperature and humidity the turnips should keep until early spring. Indoor and outdoor storage facilities are equally successful.

the only equipment needed is a container and some sand, both of which can usually be found around most houses. Indoor gardening, especially under lights, requires more of an investment if lamps and fixtures must be purchased. Cold frame gardening only calls for the materials to build the cold frame. Greenhouse gardening, obviously, is the most expensive of all if the greenhouse must be built.

For the tightest of budgets—and/or the tiniest of apartments with minimal growing space—there are seed and bean sprouts that will add fresh flavor, variety and improved nutrition to meals for the cost of the seeds and beans alone. These are relatively inexpensive, take only a few days to grow, and provide a volume of sprouts that would cost several times the price of the seeds—assuming you could find the few scattered markets that sell fresh sprouts. As an added bonus, seed and bean sprouts rank near the top of the list in nutritional value. All that is needed to grow them are containers (glass jars will do), the seeds or beans, water, and covers made of cheesecloth or light screening. No soil, fertilizer or cultivation is required.

At the other end of the winter growing scale, a greenhouse can have a sophisticated system of temperature, humidity, ventilation and light controls. The more complex the design, the more expensive the initial investment and the continuing operational costs. There is virtually no limit to the money that can be invested in a greenhouse operation, although there is a point beyond which the returns will probably no longer justify the investment.

But there is really no need to spend a great amount of money, or time for that matter. Winter growing is easy and will keep any table supplied with fresh produce throughout the long, cold months for a very modest investment.

Cellar Gardening

A cool cellar is an excellent environment in which to grow a number of crops during the winter months. The easiest to cultivate are vegetable greens used in salads, but other root vegetables such as rhubarb and asparagus may also be forced all through the winter in a cellar garden.

The only hazards threatening winter cellar crops are the fumes generated from gas, coal or oil furnaces. If the cellar is partitioned, or if electricity or solar energy are used to heat the house, there is no danger to the crops. But furnace fumes in amounts so small that they cannot even be smelled are enough to stunt or kill a winter cellar crop.

A cellar temperature between 50°F to 60°F is ideal for vegetables. Some plants may need darkness to grow. These can be cultivated in a corner of the cellar that has been curtained off or placed under a cardboard box in which a few holes have been punctured to provide ventilation.

Salad greens can be harvested from vegetables and roots such as beets, carrots, parsnips, turnips, rutabagas, chard, celery, parsley, cabbage, dandelion, kale, collard and kohlrabi. A mixture of these vegetables and roots can be planted at the same time in the same container to yield a variety of salad greens all winter long. Crooked or misshapen roots that would not normally be eaten will produce salad greens just as tasty as normally shaped roots.

Winter salad greens should be grown in sand that is kept moist throughout the growing. Since sand-filled containers are very heavy, the containers used should be small. An ideal container is a wooden crate used to ship apples. If apple crates are not available, other containers such as bushel baskets, galvanized buckets or

Sand-Filled Containers for Cellar Gardening

tubs, five-gallon metal pails, crocks or clay pots may be used. Drainage is not a consideration so there need be no drainage holes in the bottom. Whatever the containers, make sure they are clean (especially if they held other liquids or materials previously) and dry before placing the sand and roots in them. Do not use cardboard containers; they are not sturdy enough to hold the weight of the sand and will leak water.

Place a layer of sand several inches deep inside the container, set in the roots with the tips down and the crowns up, cover with another inch or two of sand and moisten well. Roots planted in this way will produce greens for as long as two months before giving out. Plan on several cuttings from each planting and keep several boxes producing at all times to be sure of a continuous supply. To keep the cellar clean, spread canvas under the containers or line the boxes with plastic to prevent the sand from leaking out.

Cellar-grown salad greens do not require a great deal of sunlight. The subdued light found in most cellars is generally sufficient. If your house has no cellar, set the containers in a seldom used room or on a porch with the proper temperature.

Mushrooms are another cellar garden favorite. Instructions for growing them are given in the directory.

Indoor Growing Conditions

At any time of the year, successful gardening depends on providing plants with the proper amount of light, a soil conducive to good plant growth, a temperature and humidity range beneficial to the plant, fertilizer of the right type, and freedom from insects and other plant pests.

Outdoors during the summer, most of these conditions are pretty much dictated by the particular area in which you live. The only way to increase a garden's success is to plant crop varieties better adapted to your local conditions.

Indoors during the winter, however, these conditions can all be controlled to greater or lesser degrees. For example, vegetables that do not take to a sandy soil can be planted in containers with a heavier soil mixture. The hours of light indoors can be manipulated by using artificial lights that provide the right combination of colors needed to stimulate photosynthesis. Within limits, temperature and humidity can be varied to provide the proper atmosphere for healthy plant growth. Fertilizers are basically the same indoors or out, summer or winter. But an added advantage of indoor winter gardening is that with proper precautions plant pests are minimal.

Some elements of successful winter gardening vary with the plant or vegetable being grown. These differences are noted in the directory instructions for each plant. The general growing conditions are discussed below.

Soil—For most indoor gardening, potting soil available from nurseries, plant stores and garden supply shops serves well. Potting soil has the proper mixture of different soil elements and comes already mixed and sterilized. Under most circumstances outdoor soil should not be used. Its major disadvantage is that bringing it indoors also brings along bacteria, fungus, insects and their larva and other pests, as well as weed seeds—all of which can do great damage to an indoor garden and are stimulated by the favorable conditions indoors.

Potting soil can be modified to suit the particular needs of each plant. For plants requiring better drainage add sand, vermiculite, perlite, leaf mold or peat moss. Some plants prefer soils that are more acidic, others like soils that are more alkaline. Soil can be made more acidic by adding vinegar water, peat moss, aluminum sulphate, tea leaves or coffee grounds. Soil can be made more alkaline by the addition of some form of lime, bone meal, egg shells, cigarette or fireplace ashes, or animal manure.

Inexpensive soil test kits are available to test the pH of a soil. The pH factor is rated on a scale of 1 to 15. Any number higher than 7 indicates alkaline soil; any number lower than 7 indicates acidic soil. Over a period of time soils tend to become slightly acidic depending on the pH of the water used. Plants that are sensitive to the pH factor are noted in the directory. If they are not growing as well as expected, check the pH of the soil and adjust it as needed.

Temperature—The ideal growing temperature should range during the day between 68°F and 80°F—preferably between 70°F and 75°F. At night the temperature should drop approximately ten degrees from the day's high. The drop at night aids the growing process and increases the harvest. Plants grown in an indoor winter garden are native to the outdoors where the temperature drops naturally at night.

Humidity—The humidity indoors during the winter will fall to very low levels if the house is not humidified through the heating system. A relative humidity of 20% to 30% is not unusual in a house that is not well insulated and airtight. Plants, however, need a relative humidity of at least 40% and are more comfortable with a humidity in the 50% to 60% range. A low relative humidity causes the soil to dry out rapidly and the surrounding air leaches moisture from the plants. A low relative humidity can cause plant foliage to dry up even though the plant is kept watered.

There are a number of ways to increase the relative humidity in the indoor winter gardening area. The least expensive way is to use plastic trays sold in garden supply shops for growing house plants indoors. Spread a layer of perlite or rock chips at least an inch deep in the bottom of the plastic trays and pour in water regularly. The water level should remain below the top of the perlite or rock chips so that the water does not come in contact with the bottom of the plant container. Plants should not sit in water because the roots will rot. The perlite or rock chips support the container above the water level. The water evaporates and increases the humidity where it is most needed—around the plant.

Other methods to increase humidity include placing glasses of water between the plant containers. A cool air humidifier can be purchased to control the humidity. It should be one with a humidistat that controls the fan and keeps the room at a constant humidity level. Too much humidity in a room can damage the walls. A good way to determine if the humidity level is approximately correct is to watch for condensation on the windows. When droplets appear on the windows the humidity level has been raised too high and damage may be done to the house.

Another way to increase humidity and tend to the breathing of the plant's leaves at the same time is to mist the plants daily. Any kind of bottle with a spray mist attachment may be used—window spray bottles, hair setting lotion bottles, etc. Just be sure they are thoroughly cleaned before using them to mist plants. Plastic and brass misters are also available from plant and garden shops; the plastic containers generally have a larger capacity than the brass atomizers. Misting increases the level of moisture in the air surrounding the indoor garden and also provides moisture for the plants to absorb through their leaves.

The most efficient way to regulate the humidity in homes heated by forced air is to attach a humidifying system to the heater which will control the relative humidity level throughout the house. Still another way to provide humidity for house plants is to grow them in the most humid rooms. The kitchen, bathroom, and laundry rooms are generally the most humid rooms in a home because water is regularly added to the air through cooking, bathing, and laundering.

Check the humidity level with a hygrometer. These inexpensive instruments are sold in hardware, garden supply and plant shops and are the only accurate way to measure the humidity.

If plants are grown in an area such as the basement where it would be impractical or very expensive to humidify the entire room, wall off the area with polyethylene plastic. Be sure to allow at least minimum ventilation and be sure the temperature is kept within the desirable range. Several fluorescent fixtures over the plants may keep the heat to the proper level.

Ventilation—Plants need fresh air just as humans do, but they should not be subjected to cold drafts which will damage their growth. One way to increase the ventilation is to install fans to pull fresh air in or exhaust stale air. Kitchen or bathroom exhaust fans can often be just enough to increase the ventilation and bring fresh air to the growing plants. Fresh air not only helps prevent disease organisms from growing on moist plant leaves, but also brings in the oxygen roots need and the carbon dioxide used by the leaves in photosynthesis.

Water—Plants grown in an indoor winter garden need particular attention paid to their water supply. Indoor plants may need more watering than the same plants would require outdoors, if for no other reason than the fact that the natural atmosphere provides most of the water for outdoor plants whereas the only water the plant receives indoors is provided by the gardener. If the relative humidity is low, the soil will dry out much more quickly than if the humidity is maintained at an appropriate level.

Unfortunately there is no uniform rule for watering plants. Much depends on soil composition, the type of container used, the relative humidity, and the kind of plant. Sandy soil drains more rapidly than soil with a higher proportion of clay. Clay pots will allow water to evaporate much more rapidly than plastic pots. Some plants prefer to be dry; some prefer soil that is always moist. Much of this is trial and error. While some people are afraid of underwatering, others are fearful of drowning their plants. Both are possibilities, but the fear of underwatering has tended to make gardeners overwater.

When soil is dried out, it is lighter in color, feels dry and weighs less. If the soil is dry to a depth of at least 1/2 inch, the plant needs to be watered. Give the soil a good soaking to be sure that the water reaches the bottom of the container where all the roots can obtain it. It is best to water plants from the top—this is the way they receive their moisture when planted outdoors. Water in the morning while the temperature is rising. When watered during the day, plants tend to wilt from the heat. If watered at night, plants may develop diseases.

If the tap water is chlorinated, allow it to sit for 24 hours during which time much of the chlorine evaporates. The water should be at least the temperature in the house and perhaps a bit more tepid. Never water plants with very hot or very cold water, both of which will damage the roots.

Fertilizers and Pesticides—Plants grown indoors are fertilized and treated for pests with the same kinds of fertilizers and pesticides used for outdoor plants, and in the same general proportions, although indoor plants sometimes require more frequent fertilization because they do not have access to all the natural nutrients contained in outdoor soil. On the positive side, they are usually less bothered by pests than outdoor plants because they are grown in sterilized soil.

While it is difficult to generalize about fertilization, most food-producing winter grown plants should be fed approximately every ten days. Use a balanced water-soluble fertilizer like Peter's Special, which has an 18-18-18 composition (nitrogen-phosphorus-potash), and apply it with the regular waterings. Nitrogen is the nutrient that promotes healthy leaf growth. Phosphorus helps plants mature their crops. Potash produces strong stems and good root growth.

Leaf vegetables may prefer a fertilizer with a higher concentration of nitrogen; root vegetables may prefer one with a higher concentration of potash. The most sensible approach is to adjust the feedings and the fertilizer composition to suit the particular needs of the crops being cultivated, which are described in the individual articles in the directory. In any case, *do not overfertilize.* Underfeeding will stunt plants; overfeeding will kill them.

Light—Perhaps no single condition requires more care and attention from the indoor gardener than light. Without sufficient light, and the right *kind* of light, there

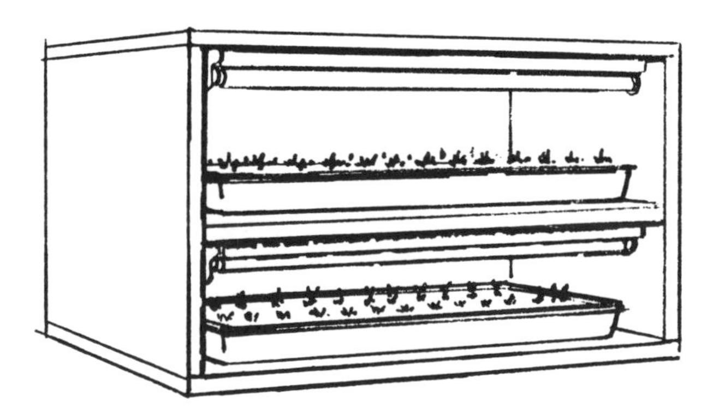

Bookcase Unit with Lights for Indoor Growing

cannot be proper plant growth. This is particularly true of many plants grown for food.

During the summer, the sun shines long enough to stimulate plant growth and crop production. In winter, however, the sun does not shine long enough (except in certain southern and tropical regions) to provide the 14 or more hours of light most plants need every day for healthy growth and bountiful crops. A potted plant on a windowsill that flourishes naturally in summer may wither in the same spot during the winter for lack of sufficient light.

To compensate for the limited winter sunlight, a variety of artificial lights have been developed. These lights and their application to indoor plant growth are an important topic in themselves—a topic that is considered in the next subsection.

Growing Under Lights

In recent years scientific studies have been conducted to determine the type of growing light that will best substitute for natural sunlight. Plants assimilate light through photosynthesis, a process in which light rays combine with carbon dioxide and water to manufacture sugars and starches. Not all colors in the spectrum of light rays, however, stimulate the process equally well.

Scientists have found that red and blue light rays have the most significant effect on plant flowering and growth. Red light rays alone make plants grow rapidly, often becoming stalky and tall. Blue light rays alone produce plants with dark green leaves and thick, short stalks. A correctly balanced combination of red and blue light rays will bring proper plant growth. This combination is most often found in fluorescent lights.

A wide variety of fluorescent lights are now being marketed. Each type offers different advantages. Standard fluorescent bulbs come under names such as Natural White, Cool White, Deluxe Cool White, Warm White and Daylight. The designations do not refer to the heat the bulbs generate, but rather the combination of reds, blues and other colors that each bulb produces. These bulbs are most regularly used to light offices, homes and public places. They are designed more for what they contribute visually to a locale than what they will do to promote the growth of plants. Still, indoor gardeners have used standard fluorescent bulbs in different combinations as a successful substitute for natural sunlight. These combinations include: Warm White/Cool White, Warm White/Daylight, Warm White/Natural White, Daylight/Natural White, Deluxe Cool White/Daylight, Deluxe Cool White/Natural White. Bulbs in combinations are better than just one type of lamp. Cool White bulbs have also been used together with incandescent bulbs to provide light for growing vegetables. Incandescent bulbs, which can be mounted in special fixtures with sockets for both fluorescent and incandescent bulbs, provide the red light rays (as do standard fluorescent bulbs with "warm" designations) and the Cool White fluorescents supply the blue light rays. Incandescent bulbs, however, generate much more heat and use more electricity than comparable fluorescent lamps. Between 15 and 30 watts of incandescent light combined with 80 to 100 watts of Cool White fluorescent light strike a proper balance for vegetable growing.

Two classes of plant growth lamps are available now, and a third type of fluorescent light, somewhere between the standard and plant growth types, is also being sold. The first class of plant growth lights reaches farther into the red area of the spectrum and has more successfully stimulated plant growth than the other types of fluorescent lights. Gro-Lux Wide Spectrum, manufactured by Sylvania, Agro-Lite made by Westinghouse, and Vitalux A which is made in Japan, can all be used without incandescent lights. All of these lamps cast a very pale pink glow that helps promote the most natural appearance in plants growing indoors.

The second type of plant growth light, which was

actually invented first and costs more than any other types of fluorescent lamps used in gardening, are best used in combination with incandescent lights, although they can be used alone. Gro-Lux made by Sylvania, Plant-Grow made by Westinghouse, Vitalux and Plant-Lux made in Japan, Plant-Light made by General Electric, and Osram-L-Fluora made in Germany are all similar in the color composition of the light rays they generate. They have all been used, with or without incandescents, to stimulate plants and vegetables to grow healthily indoors.

Natur-escent and Vita-Lite, both manufactured by Duro-Lite, arc fluorescent lamps with higher proportions of green wavelengths, the area of the spectrum that plants reflect, and were not designed to be used solely to help plant growth indoors. Vita-Lite has a higher amount of ultra-violet emissions than Natur-escent and can help stimulate growth in herbs and leaf vegetables.

There are other types of bulbs that are now mostly experimental. These include the high intensity discharge bulbs such as mercury, sodium, xenon, and metal halide lamps. They require special installations and operating know-how which, combined with the greater amount of heat they discharge, make them impractical at the moment for most indoor home gardening operations.

The bulb or combination of bulbs to use for indoor growing is a question of cost versus quality. The least expensive lamps are the standard fluorescents, but they also are the least effective. The lamps that have the greatest effect on winter plant growth are the wide-spectrum bulbs, which cost more. To make plants look their best, Vita-Lite and Naturescent are good choices since they have less tendency to change the natural coloring of the plants.

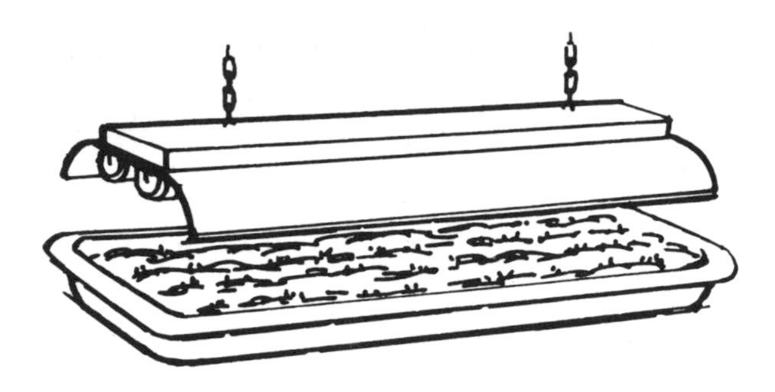

Seedlings Started under a Light Unit

Growth lamps come in a variety of sizes from 14-watt bulbs that fit into 15-inch fixtures to 215-watt bulbs that fit into 96-inch fixtures. The fixture size and bulb wattage should be matched to the desired growing space, taking into account the fact that the intensity of the light diminishes near the ends of the bulbs. A fixture of two or four 215-watt bulbs would be much larger than necessary for a small herb garden, which would do perfectly well with the 40-watt, 48-inch fixtures. Larger plants and vegetables would do better with larger fixtures.

Light fixtures should be installed in a way that will allow them to be positioned close to the top of the plant during all stages of growth. If seeds are germinated

Lighting Fixture Adjusted for Growing Plants

under the lights, the fixtures will have to be lowered within two to five inches of the containers. As the plants grow, the lights will have to be raised, so that by the time the plants are nearly full grown the lamps are six to twelve inches above the plant tops. The fixtures can be hung on chains with hooks or attached to pulleys that will permit them to be raised and lowered as needed.

To receive the greatest benefit from lights, arrange the indoor garden so that as much light as possible is reflected back onto the plants. While not absolutely necessary, it is best to paint the nearby walls and ceiling with a flat white paint to help fight reflection. Mirrors may also be placed on all sides of the plants. Use shiny aluminum pie tins or cooky sheets as saucers under the plants to reflect the light up to the bottom leaves. To clean the lamps, remove them from the fixtures and use window cleaning liquid all around the bulbs to remove the dirt. Unlike incandescent bulbs that burn out all at once, fluorescent lamps deteriorate over a period of time. When the light level becomes noticeably lower, it is time to replace the bulb. Fluorescents, however, do not have to be replaced as frequently as incandescents. Fluorescent bulbs generally last 15 times longer than incandescents.

While plants growing indoors need the proper amount of light every day, they also need a nightly rest from the light. Attach a 24-hour timer to each fixture that will automatically turn it on and off, relieving you of that task and guaranteeing uniform periods of light and darkness each day. Most winter garden plants require 12 to 16 hours of light daily. Experiment with the hours of light until the maximum benefit is achieved. It is possible to have too much light, either by placing the lamps too close to the plants or by leaving them on too long each day. And, of course, it is also possible to have too little light by placing the lamps too far away from the plants or by not turning the lights on long enough each day. For best results in the short daylight hours of midwinter, divide the extra period of artificial light evenly between the morning and evening hours.

Cold Frames and Hot Frames

Outdoor gardening does not necessarily have to end when the long summer days shorten as winter approaches. It can continue through the winter in cold frames and hot frames—devices that protect crops from frigid air while at the same time trapping the sun's heat (and generating some of their own in the case of hot frames) to promote plant growth.

The main difference between a cold frame and a hot frame is the heat source. A cold frame has no heat beyond that supplied by the sun. A hot frame uses electricity, fermenting hay or fresh horse manure to heat soil from below. (The choice between a cold frame or hot frame is determined by what is to be grown, what the weather is like during the winter and whether the crop is to be cultivated through the coldest part of winter.

In many areas of the country a crop such as lettuce, for example, can be grown either outdoors in a garden or in a cold frame for 300 days a year, with the gardening halted in only the two most severe winter months. A hot frame makes it possible to continue growing such crops even through those winter months by heating the soil—at an additional cost for heat, of course.

Both cold frames and hot frames should be located in a place that receives the maximum amount of available sunlight, has good drainage, and is protected by other structures from the wind. They should not, of course, be situated beneath roofs where they might be subjected to run-offs from rain and melting snow. The ideal location is against the foundation of a garage or house, facing south, with hedges or some sort of obstruction near enough to reduce wind but not cast a shadow over the frame. Wind protection is the least important considera-tion because leaves, straw, or other mulch materials can

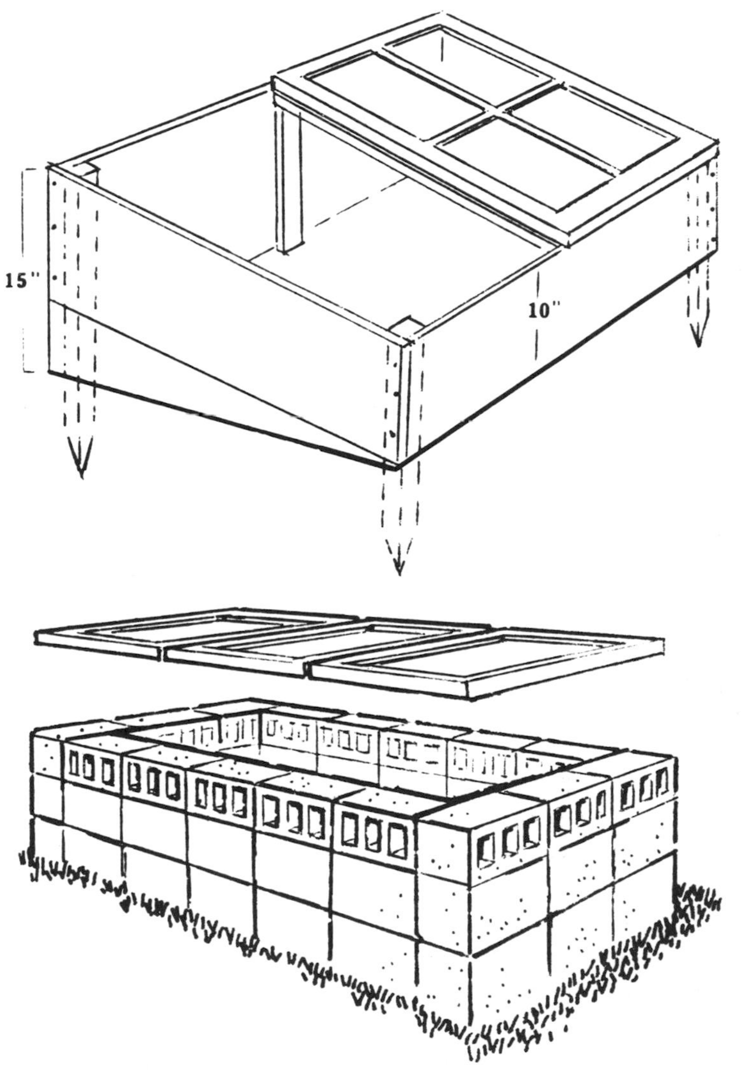

Models of Wood and Concrete Block Cold Frames

be piled against the outside of the frame during the cold, windy months to provide additional insulation.

The basic structural design of both cold frames and hot frames is the same. A window or storm sash is placed on a frame and hinged so that it can be opened and closed to adjust the inside temperature. The back of the frame is higher than the front to maximize the amount of sunlight that can enter. Convenience and availability are the most important factors in planning the dimensions of a frame. The height to which the plants will grow should also be considered. Construction of a typical frame is shown in the illustration. The frame is built of 2 x 10 lumber cut and fitted to the proper length, as determined by the size of the window. The corner posts should either be cut long enough to be sunk about six inches into the ground to anchor the frame, or the entire frame should be set at least three inches into the ground.

In parts of the country where termites are a threat, or where greater insulation within the frame itself is desired, concrete blocks may be used to construct the frame. As shown in the illustration, the top course of blocks should be laid without mortar so that they can be turned on their sides to allow ventilation.

The soil for a cold frame can be any good, easily crumbled gardening soil. If the available soil is not of sufficient quality, a good growing soil can be mixed from two parts garden loam, one part fine, sharp sand and one part compost or leaf mold. If necessary, adjustments can be made in the soil's pH factor by adding acidic or alkaline materials.

A cold frame can be converted into a hot frame by adding a source of heat. A natural source of heat is fresh horse manure. A mixture of two parts manure, one part of the horse's bedding litter (straw, hay, sawdust or sphagnum may be substituted), combined with one part fresh leaves will extend the period over which the manure will provide heat. Mix all of the materials well in a pile and water thoroughly. After a few days, when the pile is steaming, mix it up well once again. After another week or so, the material is ready to use. Excavate two to three feet of the earth inside the frame. Spread the material to a depth of six inches at a time and pack it down between each spreading. Allow six inches at the top for the soil in which the crop is to be planted. Seeds can be sown when the temperature of the soil in the hot frame has dropped to 90°F.

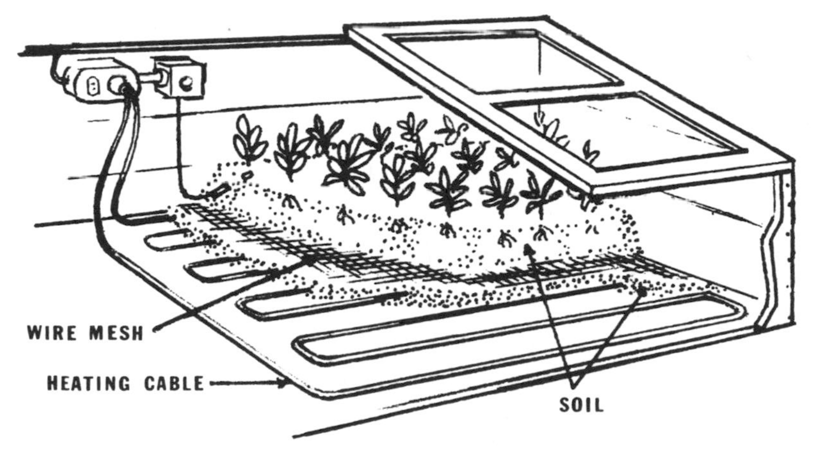

Model of a Hot Frame

A better although more expensive way to add heat is to use electric heating coils which are available from nurseries or mail-order garden supply houses. The cables are low wattage—usually 3 1/2 watts per foot—and 36 feet of cable should be enough to heat a frame with approximately 20 square feet of growing space. Excavate six or eight inches of soil from the frame. Lay the cable in the bottom of the frame as shown in the illustration and cover it with about an inch of soil. Lay a wire mesh screen over the soil to prevent the roots of the growing plants from being damaged through direct contact with the hot cables. Cover the screen with the top six inches of soil in which the crops are to be planted.

The thermostat in this kind of frame controls the heat much better than in a frame heated by organic materials. All hot and cold frames, however, should be watched constantly during moderate weather since heat and moisture will build up. Temperatures of 90°F in a cold frame are not unknown when the outside temperature is just 60°F. Regulate the temperature inside the frames by opening the windows.

Windows should be opened regularly even when there is no need to regulate the temperature. The object is to let in fresh air and control humidity, but care must be taken to be sure the crops are not exposed. If the window of the frame is closed for extended periods, the temperature will rise too high and the humidity will build up above the desired level, both of which may damage and perhaps even destroy the crops inside.

The first time a crop is grown in a cold frame—especially if it is in the worst of winter—the temperature inside the frame should be closely watched to see that it does not drop to levels the plant will not tolerate. If a hard frost is expected, mound leaves or other mulch materials outside the frame. If that is not enough to keep the crops protected, mulch inside the frame as well. In addition, burlap bags, blankets or other closely matted materials can be placed over the window during the night to keep the day's heat inside.

If too much sunlight enters during the day, but the temperature outside is too low to open the window for any length of time, place several layers of cheesecloth over the window to reduce the light.

Water crops carefully in both cold and hot frames. The soil should be kept moist but should not be overwatered because the frame tends to retain moisture and slows the rate at which water normally evaporates from the soil. Water in midmorning so that the excess can dry off by the time the window is lowered in late afternoon, before the temperature begins to drop for the night.

To germinate seeds in a cold frame and get an early start on spring planting, divide the seeds into two groups, those that require low temperatures to germinate and those that require warmer temperatures. Vegetables that can withstand cooler temperatures—celery, broccoli, cabbage, cauliflower and lettuce, for example—should be planted the first week in April. Vegetables that do best in warmer temperatures—cucumbers, tomatoes, peppers, eggplant, summer squash, etc.—should be planted toward the middle or end of April. The frame can be divided into two sections to accomodate the separate plantings.

Greenhouse Gardening

Winter gardening in a greenhouse is economically worthwhile only if the greenhouse already exists. Even if crops with a very high yield from a small amount of space are planned, the cost of acquiring or building a greenhouse solely for winter gardening is prohibitive. It would be cheaper to grow the crops indoors where little or no additional cost is required to heat the growing area.

An already established greenhouse, one that provides an intermediate climate where the temperature never drops below 55°F, will provide additional benefit if used during the winter to cultivate crops that produce a high yield in a small amount of space. Vegetables such as tomatoes, cucumbers, lettuce, radishes, eggplant, onions, carrots, peppers, beets, green beans, asparagus, rhubarb, endive, turnip greens, Chinese cabbage and parsley are the best candidates for winter greenhouse gardening.

Almost any vegetable crop can be grown in a greenhouse if the proper heat, light, ventilation and humidity are maintained. But produce that grows outside in summer and can be efficiently stored during the winter, or crops that require a large amount of space, should not be attempted in a winter greenhouse garden because the cost of maintaining the proper climate is far greater than the yields from the produce.

Temperature control in the greenhouse is difficult if different crops that prefer different temperatures are attempted at the same time. For example, carrots, lettuce, radishes and onions will grow best with a minimum temperature between 40°F and 50°F. The minimum temperature for tomatoes, however, is 60°F and for cucumbers 70°F.

One way to solve the problem is to regulate the greenhouse temperature near the middle range of the minimums required by the various crops. If all of the above crops are to be grown at the same time, set the greenhouse thermostat for a 60°F temperature at plant level. Then plant the cucumbers in containers with electric heating cables set in the soil. The cables should be attached to a thermostat that prevents the soil temperature from rising above 85°F to 90°F.

The greenhouse should be ventilated in the winter by exhaust fans or automatic vent openers controlled by a thermostat. The thermostat should be centrally located in a shady area of the greenhouse and set to turn on the ventilating system when the temperature reaches 70°F to 75°F.

If good quality soil is used, one specially prepared

and not just brought in from outdoors, there will be little or no problem with weeds, insects or other pests. Any problems that do arise in the greenhouse can be treated in the same way as they would be with crops planted outdoors. Likewise, the fertilizers used indoors are the same fertilizers that would be used outdoors. If crops are planted in pots, however, greater care must be taken to insure that the plants are not overfertilized and damaged.

Specific instructions for different greenhouse crops mentioned above are found in the directory.

Growing Sprouts

Sprouting seeds or beans is one of the easiest ways to add freshness to a salad or meal—and one of the most economical. For example, two tablespoons of mung beans will grow a quart of sprouts in three to five days, depending on the temperature and humidity. The cost of a quart of sprouts purchased in a food store is many times that of the two tablespoons of beans.

In addition, there are remarkable nutritional values in sprouts. The vitamin C content of sprouted soybeans increases about 500 percent in the first three days of sprouting. Many sprouted seeds show similar increases in the B vitamins and vitamin E. Starch content is reduced because the starch is converted to sugar, and protein levels remain fairly high.

The most popular seeds, beans and grains for sprouting are alfalfa, mung beans, soybeans, lentils, garbanzos or chickpeas, watercress, sunflower seeds, wheat, flax and watercress. Other grains, beans and seeds that can be sprouted include sesame, radish, mustard, red clover, fenugreek, rye, oats, corn, barley, green peas, lima beans, navy beans, marrow beans, kidney beans, pinto beans, cranberry beans, fava beans, parsley and millet. *Do not sprout tomato seeds or potatoes because their sprouts are poisonous.*

Buy only seeds or whole grains that have not been chemically treated against insects or other diseases and are sold for eating or sprouting. They are available in health food stores, groceries and supermarkets, as well as from some nurseries, garden supply shops and through the mail from growers and suppliers. Store them in a cool, dry place.

The equipment needed for sprouting is readily available in most homes. For small quantities of sprouts any wide-mouthed quart jar will suffice. Empty mayonnaise jars are ideal. Inexpensive sprouting trays and containers are also sold in most health food stores.

Soak two to three tablespoons of seed or grain overnight. Cover the top of the jar with cheesecloth, nylon or wire mesh. The size of the mesh is determined by the size of the seed or grain being sprouted. Attach the mesh cover with a rubber band, string, or a piece of wire. Drain the soaking water (it can be reserved for use in soups) and place the container in a dark, humid place. Rinse the sprouting seeds or grains several times a day with lukewarm water and drain off all the excess. Depending on the temperature and humidity and the type of seed or grain being sprouted, the sprouts will be ready to eat or use in cooking in three to six days. Rinse and drain sprouts and if desired remove any hulls that have not floated off during sprouting. Any sprouts that are not used immediately should be refrigerated; they will keep for several days.

Sprouts can be eaten raw as a snack, in salads, sandwiches, or added to soups, casseroles, stews, meat loaves, souffles, vegetables, omelets, breads, muffins and waffles. They can be cooked by themselves, but smaller sprouts should not be heated more than three minutes and larger sprouts more than five minutes. Sprouts should be added to cooked foods only at the last minute to retain their nutritional value. Put the sprouts through a food grinder before using them in cooked cereals, breads, and other baked goods.

The approximate length of sprouts varies with the grain, bean or seed being sprouted. Following are the lengths these seeds and grains reach when the sprouts are ready for harvest: mung beans – one and one-half to three inches; alfalfa – one to two inches; wheat – when the sprout is the length of the seed; chick-peas – 1/2 to 3/4 inch; lentil – one inch; flax – 3/4 inch; soybean – one-half to two inches; green pea – two inches; sesame and unhulled sunflower seeds – as soon as sprouts can be seen. Most sprouts are not as tasty if allowed to grow too long.

DIRECTORY

An alphabetical listing of the fruits, vegetables and herbs suited for winter growing, along with the conditions and methods for cultivating each

Artichoke

Globe artichokes are space consuming plants and need an area where they can have about a five-foot spread and a three-foot height. The plants cannot tolerate a frost and must be grown in a well-insulated hot frame or greenhouse where the temperature is always kept above freezing. They may also be grown in large pots on a winterized sunporch that is free of drafts.

Artichokes are regarded as a winter vegetable, since they produce during the months of December through April. Outdoors, they like a frost-free area in winter and one that is cool and foggy in summer. The closer the conditions for indoor growing approximate these, the better the chances of success with the plants.

Artichokes cannot be propagated from seed. Buy starter plants from a nursery and plant them in a spot where they will not have to be moved. The plants are temperamental, difficult to grow, and do not take well to transplanting or shifts in location. They like a rich soil and are heavy feeders throughout the growing period, so they should be well fertilized. They need a good supply of water and fare best in a humid atmosphere; an occasional misting can be helpful.

Harvest artichoke globes before the outer leaves become woody. They are treasured for the tender leaves and if not harvested before the leaves harden they will be tough and stringy. The globe artichoke is not related to the Jerusalem Artichoke, which is not an artichoke and doesn't come from Jerusalem. This tuber is related to the American sunflower and is grown like a potato. It is not well suited for winter gardening.

Asparagus

While asparagus can be grown from seed the best method is to begin with already established year-old root crowns which can either be dug up from a summer garden or purchased from a nursery ready to plant. The roots fare best in a rich, well-drained soil with a good amount of well-rotted manure mixed in. Top dress the plantings with a liberal application of commercial fertilizer and water well throughout the growing.

For cellar gardening, dig the roots in the fall and transplant them into boxes of earth. Grow the plants in a light, heated place and water well. To space out the harvest, hold some of the roots in a cool place and plant them at different intervals.

The roots can also be grown in pots, under lights, one

root to a six-inch pot, or two roots in a twelve-inch pot. Do not plant the roots too deeply; they should have no more than two to three inches of soil over them. To force asparagus roots in a greenhouse, pack them under a three-inch layer of damp peat moss and water well. Harvest spears for as long as eight weeks and then replant the roots outdoors so they can rest to provide next year's crop. The greenhouse temperature should not be less than 60°F.

No matter where asparagus is grown the plants need enough water or the spears will not be full and succulent. The plant also needs its summer rest and top growth or the crop will not come back at its best.

Bananas

Bananas can be grown indoors or in a greenhouse. Order the Dwarf Banana from a nursery; it grows to a height of four to six feet. It should be planted in a tub with a porous, rich soil that contains a good amount of organic matter—compost, rotted leaves, etc. The tub should provide ample drainage so the plant does not drown. Place the banana in full sunlight, protected from drafts, and do not let the temperature drop below 60°F. Blackened leaves are a sign of overwatering.

Spray the plant leaves with water once or twice a week to maintain proper humidity. To provide the moist atmosphere in which bananas thrive, it is a good idea on occasions to spray the plant and cover it overnight with a large plastic bag. Remove the bag in the morning before the sun heats it. When the plant reaches full growth, it may be necessary to make a bag out of a 9 x 12-foot plastic sheet.

After the plant fruits, it will wither, die off and then send up new growth that will bear fruit the next year. Do not place near other plants because the fruits give off ethylene, a gas deadly to many plants.

Basil

Basil is best grown indoors under light or in a pot on a windowsill. Sow the seed in a small pot with a propagating medium such as Pro-Mix, or in a Jiffy 7 pellet. Place the containers under fluorescent lights set five inches above the container tops. Set a timer for 14 hours of light each day. Do not water seedlings too much.

When roots begin to show through the sides of a Jiffy 7 pellet, or the true leaves appear above the seed leaves, place the plant in a three-inch pot filled with sterile potting soil that is not too rich. Keep the fluorescent lights five inches above the top of the plant and leave the lights on for 14 hours a day. Pinch off the top leaves to stimulate a compact bushy plant rather than a long spindly one; the former produce more flavorful leaves with more essential oils. Use a liquid fertilizer once every 10 to 14 days.

Harvest basil before it blooms and goes to seed because once the plant flowers, the amount of aromatic oils in the leaves diminishes and the herb will be less flavorful. Cut off the main stem at a height that will leave at least one node and two young shoots. Fertilize lightly after cutting to help stimulate new growth. In another two to three weeks, the plants should be ready to harvest again. Use fresh basil as it is harvested, or dry according to the instructions in the drying directory.

Bay Laurel

The best way to grow a bay laurel is to buy a starter plant from a nursery. Seeds are very difficult to germinate and cuttings easily require six months to root. The plant, native to the Mediterranean area, dislikes dry indoor heat, so measures will have to be taken to provide the level of humidity in which it fares best. The permanent growing soil must be rich and well drained. To each quart of sterile potting soil, add 1/2 cup each of dried cow manure, peat moss, leaf mold and sand. The plant is quite large when fully mature and is often grown in tub containers.

Herb growers differ on the type of light bay laurel requires. Some say direct overhead sunlight, as in a greenhouse, is best. Others say it needs indirect sunlight and that the light of the noonday sun will burn the leaves. There is no disagreement, however, that the plant must be protected from frost and icy winds.

Cut the bay leaves from the side of the plant rather than from the top. They can be used either fresh or dried.

Beans, Green

Green snap beans are very good for greenhouse growing. They require temperatures at least 60°F but not higher than 85°F. Sow seeds in peat pots in February or March and transplant into beds when the seedlings are four to six inches tall. Plant the seedlings at least 18 inches apart. Train the pole-type plants on stakes or rope. Bush-type beans can also be grown but require more light. Sow the seeds directly into the beds in rows that are 12 inches apart with two to three inches between each seed.

Both bush and pole beans can be grown indoors under lights with 14 hours of light each day. Keep the lights positioned five inches above the plant tops and provide stakes for the pole beans to climb.

Beans do not like a mucky or clay soil. They fare best in a well-drained warm garden soil, sandy rather than heavy, which contains moderate amounts of humus and manure. Outdoors, beans draw nitrogen from the air as they grow; since this may be less available indoors, it will have to be provided through regular fertilization.

Beets

Beets are a double winter crop—they can be grown for the greens, a fine addition to salads, or for the beet roots.

To grow beets for greens, plant otherwise useless or misshapen roots in a sand-filled box as described in the section on cellar gardening. Keep the sand moist, occasionally adding a bit of liquid fertilizer, and the roots will provide ample greens for table use. Beets can also be grown for greens from seeds. Sow the seeds thickly together and cover with 1/2 inch of soil. Fertilize and water the seeds generously to promote rapid growth. There is no need to thin the seedlings since they are being grown for the greens only. Harvest the tops when they are five to eight inches tall.

Beets may be grown for the roots indoors under lights or in the greenhouse. The smaller, early maturing varieties are best for this purpose. Sow the seeds about three inches apart and cover with 1/2 inch of soil. After the seeds germinate and the seedlings appear, thin them to about 12 to 14 plants per square foot. Greens of plants that have been thinned out can be used in salads.

Harvest beets when the roots are about an inch wide across the tops. While beets can grow up to three inches across the tops, the smaller beets are tastier eating. Make several successive plantings to provide beets at a steady rate rather than all at once.

Beets will grow in almost every type of soil but do best in a rich sandy loam that contains a good amount of well-rotted cow or horse manure. They are heavy feeders and should be fertilized regularly.

Borage

Borage can be used in a salad or made into a tea. Plant the seed in a Jiffy 7 pellet or in a propagating soil. Place the containers under fluorescent lights that are timed for 14 hours a day. Transfer the plants to pots filled with sterile potting soil with added bone meal or dried manure. Borage likes a good deal of water and needs the extra food the bone meal or manure provides. Replant into larger pots as the plant grows. Borage prefers a well-drained friable soil that is not too rich.

If the leaves are to be used in salads, harvest by picking stems from the center of the plant to discourage flowering that will reduce the amount of aromatic oils in the leaves and make the herb less flavorful. If the plant is to be used for tea, allow it to flower and then use leaves and flowers in the tea.

Broccoli

Broccoli is best grown in a cool, moist greenhouse. Use the sprouting, rather than the heading-type broccoli because the latter takes much longer to develop.

Buy started plants from a nursery or start the seeds in peat pots indoors. Broccoli plants require five or six square inches per plant. They are remarkably tolerant of different types of soil as long as they are well watered. They are moderately rich feeders so the growing soil should contain good amounts of fertilizer, preferably well-rotted dried manure. A light top dressing of nitrate of soda is also helpful.

Harvest broccoli before the buds open or the heads will become rubbery. After the main head is cut, a few small shoots will develop from the axils of the remaining leaves. These will produce flower buds that are also edible.

Brussels Sprouts

Brussels sprouts are one of a few vegetables that will grow well into winter and provide a hearty crop outdoors as late as December or January. Some gardeners insist that a few good frosts actually improve their flavor.

Brussels sprouts do not like excessive heat or dryness, and adequate moisture is necessary for their cultivation. They do best in a soil to which well-rotted manure or a good commercial fertilizer has been added. The plants may also have a light top dressing of nitrate of soda. Too rich soil or too much fertilization will cause the plants to become leggy.

Buy plants from a garden store or start the seeds indoors in peat pots. Brussels sprouts require five or six square inches of space per plant. Sow the seeds in late spring and transplant them outdoors in late summer. Set the plants 16 to 20 inches apart in rows with the same distance between them. Break off the lower leaves in October to force the growth into the buds. Harvest the lower buds first. In southern areas, Brussels sprouts may be harvested from November through March. In more northern areas, December or January seem to be the latest harvest months.

Cabbage Roots

Cabbage roots can be dug out of the ground after the cabbage has been harvested and used in a sand-filled box, described in the section on cellar gardens, to provide salad greens all winter long. The cabbage itself stores so well there is no need for it to be winter grown.

Carrots

Carrots are one of the winter gardener's most versatile crops. They can be left in the ground, used in a cellar garden to produce greens, and grown indoors under lights or in a greenhouse.

Gardeners have recently discovered that carrots can be left in the ground all winter long and even when frozen will produce a tender crop. For best results, the carrots should be mulched to protect them from the most severe winter weather. Misshapen carrots can be placed in a sand-filled box to produce salad greens, as described in the section on cellar gardening.

For the indoor or greenhouse garden, the best varieties of carrots are the midgets. These carrots are usually three to five inches long when fully grown and take anywhere from 50 to 70 days to reach maturity. Some good varieties to winter grow are Tiny Sweet, Little Finger, Short 'N Sweet, Red Apple, Gourmet Parisienne, Gold Nugget, Sugarstick, Baby-Finger Nantes, Sucram and Midget.

For the fastest and best root development, carrots prefer a rich, friable soil that drains well. Use one of the sterilized potting soils or a mix such as Pro-Mix B and add 1/4 cup of sand to each quart of soil or mix. Be sure the containers or pots are deep enough to accomodate the length of the carrots grown. Sow seeds closely and thin them out as they grow so that the roots are not touching. Sow the seeds approximately three or four to the inch, cover with 1/4 inch of soil and then thin the plants to about 3/4 inch apart. As the carrots grow, pull every other small one for table use and leave the rest 1 1/2 inches apart to grow longer. If carrots are not thinned, the crop will be all tops and no bottoms.

Growing time indoors will be longer than the time stated on seed packets, which is calculated for outdoor growing. The indoor growing time is determined by the atmospheric and light conditions and is usually at least a week longer. Carrots should be watered daily and fertilized every ten days. Grown under fluorescent lights, the lights should be on 14 hours a day or longer. Plant every three weeks to have a continuous supply.

Celery Roots

Dig up celery roots to use in a sand-filled box, described in the section on cellar gardening, to grow a supply of celery greens for salads through the winter. The celery roots referred to here are the roots of the stalk celery and are not to be confused with celeriac or celeri rave. Celeriac, cultivated specifically for the root, stores so well there is no need for it to be winter grown.

Chard

Chard, cultivated for the leafy green, is a poor shipper and not usually found in good quality or quantity in markets. It is basically a home-grown vegetable. Chard is one of the vegetables that will survive a moderate winter if left outdoors. Even if the upper growth is killed off by hard winter weather, the roots will survive and yield one of the earliest spring greens.

If chard roots are not left out for the winter, bring them in for use in a sand-filled box, described in the section on cellar gardening. Keep the sand moist, add a bit of liquid fertilizer now and then, and the roots will produce salad greens all winter.

Chard can also be cultivated in a cold frame since the frame offers enough protection, if the temperature is not allowed to drop much below freezing, to enable the plant to survive the winter while yielding a fine table crop. Seed chard in early fall and start harvesting the outer green leaves about six weeks later. Always pick the leaves from the outside and the plant will continue to provide new leaves for later harvest.

Chard will grow in almost any kind of soil, but does best in a rich sandy loam with a good amount of well-rotted cow or horse manure mixed in. The plant is a moderately heavy feeder and should be fertilized regularly. A light top dressing of nitrate of soda is also useful.

Chervil

Chervil has been called the gourmet's parsley—the flavor is more delicate and the leaves resemble a flattened parsley. The plant grows excellently indoors under light.

Sow the seed in Jiffy 7 pellets or moistened peat pots and cover with 1/4 inch of the propagating medium. Place the containers five inches under fluorescent lights that are turned on for 14 hours each day. Do not water the seedlings too much. When roots show through the pellets, transplant to three-inch pots filled with sterile potting soil that is not too rich. Keep the pots moist at all times and continue to grow them under the fluorescent lights that are placed five inches above the tops of the plants and kept on at least 14 hours a day. Leaves can be taken about six to eight weeks after seeding. Feed with a liquid fertilizer once every two weeks. For a continuous supply, seed the plant every few weeks.

Chervil prefers cool weather and can often be winter grown in a protected cold frame, where it will provide fresh leaves all winter long. Combined with tarragon and parsley, chervil is a prime ingredient in *fines herbes* used to flavor omelets. Chervil can also be chopped and used as a salad green.

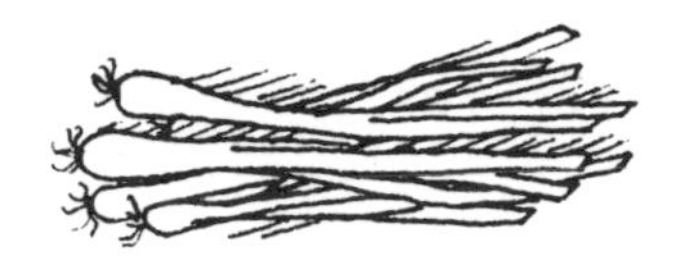

Chinese Cabbage

Chinese cabbage prefers cool weather during the greater part of its 70 to 80 day growing season. Since it suffers from transplanting (the plant tends to go to seed before the cabbage head can develop) it should be planted in the location where it will grow. Chinese cabbage can be grown indoors either in pots or containers or in the greenhouse. To grow well, the night temperatures should be around 55°F. Cold frame plantings can be started in August and continued until the first frost. Sow the seeds in rows 12 inches apart and cover with 1/2 inch of soil. When the plants are three to four inches high, thin them to about ten inches apart in each row. The thinned greens may be used in salads.

Chinese cabbage grows best in a soil to which well-rotted manure or a good commercial fertilizer has been added. The plants may also be lightly top dressed with nitrate of soda. They require a lot of moisture and should be watered each evening if possible. Harvest the heads when they mature, but don't wait too long because many varieties will go to seed very quickly if left on the plant.

Chives

Chives are wonderful plants to grow indoors or in a cold frame. The more they are harvested the better they seem to grow. While chives can be cultivated from seed, the easiest way is to divide bulbs into clumps six to twelve bulbs each. If the chives have been grown outdoors, they should be potted late in the summer. Put pot and all in the ground until after the first killing frost, then mulch around and over the pot or put it into a cold frame for about 90 days. Chives need this dormant period to rejuvenate.

Bring the pot indoors and put it on a window sill or under fluorescent lights, giving it plenty of water. By January the plant will put forth a crop to be harvested. Harvest chives by cutting off the leaves close to the crown. The crown will then grow another crop and will continue to do so as long as the plant is treated to a dormant period once a year.

Outdoors, chive spears will reach a height of 12 inches, but indoors the range is about six inches. The plant likes a moderately rich soil, plenty of water and full sun. Harvest the spears before the tiny bulbs appear at the top. These bulbs will turn into flowers and then go to seed. If seeds are desired, it is better to leave the plants outdoors and have a separate section for the chives that are to seed.

Chives ready for transplanting are often sold in nurseries and markets.

Collard Greens

Collard roots can be dug out of the ground after the greens have been harvested and used in a sand-filled box, as described in the section on cellar gardens, to provide salad greens through the winter. They can also be grown in a cold frame or under lights, using the same methods as those described for growing lettuce.

Cucumbers

Cucumbers can be grown indoors under lights or in a greenhouse. They like considerable heat, plenty of moisture, and grow best in a warm sandy loam that contains a good quantity of manure or other organic matter.

For indoor growth, select a hybrid or midget variety. Germinate the seeds in Jiffy 7 pellets under fluorescent light. When roots begin to show through, transplant the seedlings into a nine-inch pot or larger, placing two or three seedlings in each pot to simulate the mounds normally grown outdoors. Use a trellis on which to train the vines, or they will cling to nearby plants. Hanging pots work well for cucumbers since the vines can climb the pot hangers.

Grown indoors, cucumbers must be pollinated by the gardener. Outdoors, bees and certain insects will take care of pollination. But indoors the pollen must be transferred from the male flower to the female flower in order to produce the cucumbers. Male flowers are yellow blossoms about an inch across with stamen containing the pollen in the base of the petals. The female flower looks like an inch-long pickle with a little flower blooming on top of it. To pollinate the female flower, use an artist paint brush or your finger. Brush the center of the male flower to pick up the almost microscopic pollen and then brush down into the bottom of the female flower.

Hybrid cucumber plants usually have more female than male flowers to produce a larger crop of cucumbers. Any excess of females will drop off the plant as it is growing, so pollinate all the females to insure the largest crop possible. Water cucumber plants generously or the leaves will begin to wilt, signaling the need for more water. Fertilize approximately every ten days with a good liquid fertilizer.

Harvest cucumbers as soon as they are ready. Those that are left on the vine too long will cause others not to mature. When the plants mature, pick the cucumbers three or four times a week to insure that the remaining ones will grow to maturity.

Cucumbers require a greenhouse heated to a minimum temperature of 70°F. They will tolerate temperatures as low as 60°F, which is what tomatoes need, but will not grow as well as they do at higher temperatures. Cucumbers and tomatoes can be grown in the same greenhouse at the same time, but one or the other will do less well than they might if they were grown separately or at different times.

Germinate cucumbers in the greenhouse in the same manner as for indoor growing, but reduce the temperature to 70°F as soon as the seedlings appear. The plants can also be potted in the same manner and a climbing trellis should also be provided. In the greenhouse, fertilizing is tied to watering requirements. The more water required, generally the more fertilizer needed. Harvest the greenhouse cucumbers in the same manner as those grown indoors.

Dandelion Greens

Dandelion greens can be cultivated in the cellar or under lights. To use dandelions as described in the section on cellar gardening, dig the roots and place them in a sand-filled box. Keep the sand moist and the roots will produce a supply of salad greens all winter long.

Dandelion seeds can also be purchased and cultivated indoors under lights. Sow the seeds in Jiffy 7 pellets or in pots or tubs about two seeds to the inch. If sown in pellets, place in pots large enough to accomodate the long central taproot. Use sterile potting soil, adding 1/2 cup of sand for each quart of soil. Place the plants under fluorescent lights for 14 hours a day and water daily. The plants can also be grown on a window sill. If the seeds are sown in pots or tubs, thin the young plants until there is one every six inches. The thinned plant greens can be used in salads.

Date Palm

Buy dates at a natural or health food store since these dates generally still have the seeds and have not been chemically or heat treated, which will prevent germination. Notch the hard seed coating with a file. Plant the seed in a pot of peat moss mixed with sand. To germinate, the seed must have a temperature of at least 80°F and the soil should be kept moist. Once the plant has germinated and begins to grow it can be transplanted to a large pot or tub.

To produce fruit, a process that generally takes at least four or five years, both male and female plants are required. When the plants bloom, cross pollinate with a brush to set the fruit. Repeat the cross pollination a second time to be certain it takes effect.

While producing fruit is difficult, it can be done if the temperature is high enough—well into the 80's during the summer months and rarely lower than the middle

70's during the spring and fall—and the roots of the plants are kept moist but are not soaked. To ripen the fruit, the atmosphere around the plant should be fairly dry.

Dill

While dill generally will not go to seed if grown indoors, it will produce generous foliage for use as seasoning. Sow dill seeds in Jiffy 7 pellets and place under fluorescent lights for 14 hours a day. Since dill sends down a long taproot that should not be disturbed the Jiffy 7 pellets are best for indoor gardening purposes. When roots penetrate the sides of the Jiffy 7 pellet, transfer the entire pellet to a three-inch pot and fill with sterile potting soil. Feed with liquid fertilizer about every ten days and grow either under fluorescent lights turned on 14 hours a day or on a window sill.

Prune the foliage regularly to keep the plant within reasonable bounds. Because of this pruning, the plant usually will not blossom and set seeds indoors.

Eggplant

Growing eggplant indoors during the winter is difficult because the plants require plenty of heat to germinate and develop. Eggplant seeds will not germinate unless the temperature is at least 75°F and preferably between 80°F to 90°F.

Sow seeds in Jiffy 7 pellets or in peat pots filled with

either peat moss mixed with sand or one of the standard soil mixes. The best varieties for indoor growth are the midgets: Modern Midget, a purple-skinned eggplant, or Black Magic, or Golden Yellow, a yellow-skinned eggplant that grows to the size of a lemon. Plant two or three seeds in each pellet or pot and cover with 1/4 inch of soil. Eggplants take about three weeks to germinate and the process will be helped by placing the germinating containers in a place where they will get bottom heat.

After the plants have germinated, plant pellets or peat pots in pots filled with sterile potting soil to which a good quantity of rich humus has been added. Eggplant prefer soils that are reasonably rich and not too sandy. Feed plants with a good liquid fertilizer at least once every ten days. Be careful not to break the roots when placing the plant in its final growing pot.

Crops can be increased by harvesting the eggplants before they reach full size. If they are left on the plant too long they will begin to go to seed and start to shrivel. To test for ripeness, press your finger into the fruit far enough to make a slight dent. If the dent remains, the fruit is ripe.

Endive

Known as Belgian endive, French endive, or witloof chicory, this is one of the best crops for cellar gardening. While the process for producing a head may take as long as 100 days, the winter gardener is rewarded with a crop that is crisp and delicious.

Endive can be started in the ground as soon as the first hard frosts are over. Or begin cultivation in a cold frame and transfer the seedlings into the ground when they have developed three to five leaves. The root growth is principally desired, so the plants should be fertilized with a food rich in potash.

When the first chill comes into the fall air, withhold water for a few days and remove the brittle endive roots from the ground, taking care not to break them. Allow them to dry overnight and trim away top growth to the root head, leaving only a single point of brittle light green leaves. Trim off all roots below eight inches from the crown. Place the roots in a sand filled container that is two feet deep from root crown to the bottom. Keep the sand moist, occasionally adding a bit of liquid fertilizer.

Place the container in a warm, moist, dark place and in approximately four weeks the endive will be ready for the first harvest. Do not cut too deeply into the root crown and the root will provide a second and third harvest through the winter, although these later harvests will not be of the same quality as the first.

Garlic

Garlic is mainly an outdoor crop, cultivated for the bulbs that grow underground, but it can also be grown under lights. Separate individual cloves from a bulb and plant them under two inches of sterilized soil in a six-inch pot. Place the pot under fluorescent lights turned on for at least 14 hours a day. Garlic takes 90 days or more to mature. When the leaves begin to turn brown, bend them all over; after another three or four days the bulb is ready for harvest. Dry garlic bulbs in the sun for a few days to condition them and then store in a dark, dry place.

To add a light garlic taste to salads and other foods, the cloves can be grown for their greens. Using toothpicks, suspend a clove of garlic in a glass of water and it will sprout greens that can be harvested when needed.

Geraniums

Geraniums are grown indoors under light not only for the fragrances they add to the indoor air, but also for the flavoring their blossoms add to jellies. Rose geranium is the favorite for jellies.

To grow geraniums under light, root a cutting and transplant it into a pot filled with sterilized potting soil. Place the pot under fluorescent lights that are turned on at least 14 hours a day and placed five inches above the plant tops. To stimulate bushy rather than spindly growth, periodically pinch back the center leaves. Other types of geraniums can also be grown in the same manner. All but the apple geranium need to be continuously pruned.

Grapefruit

Grapefruit can be grown indoors or in a greenhouse. Buy the dwarf variety from a nursery. The plant should be in a good-sized tub with a rich, porous soil containing a considerable amount of organic matter—compost, rotted leaves, etc. The tub should provide ample drainage so the plant does not drown. Place the plant in full sunlight but free from a draft and do not let the temperature fall below 60°F. Spray the plant leaves with water once or twice a week to help provide humidity.

To encourage the humidity in which citrus fruits thrive, it is a good idea on occasions to spray the plant and cover it overnight with a large plastic bag. Remove the bag in the morning before the sun heats it. If properly cared for, the plants will blossom and produce fruit once a year.

Horseradish

The highly spiced horseradish root can be grown in a cold frame all winter long or cultivated in a pot indoors under lights. Purchase roots from a seed house and plant them in a pot or cold frame that is kept moist. Horseradish prefers a rich loam and does not do well in sandy soil. The roots can be left in the ground all winter. After the tops show a good growth, dig up the roots for use. Cut off the tops and replant these in soil. They will root and produce new plants.

Kale

Kale is another vegetable gardeners have discovered will grow outdoors through the worst of winter and provide a hearty crop in December and January. In fact, some gardeners insist that a few good frosts actually improve the flavor of kale.

Buy started plants from a garden store or start the seeds indoors in peat pots. Kale plants require about eight to ten square inches per plant. Kale can also be seeded outdoors anytime between early spring and a few weeks before the first frost. It needs time to begin growing before the frost sets in.

Kale does not like excessive heat or dryness, and adequate moisture is required for its cultivation. It does best in a soil containing good amounts of well-rotted manure or a quantity of good commercial fertilizer. The plants may also have a light top dressing of nitrate of soda.

Kale can be harvested either by cutting the entire plant or taking off the larger leaves while they are still young. Old kale is not very tasty. If the entire plant is cut, dig up the roots and place in a sand-filled box as described in the section on cellar gardening to grow a winter-long supply of salad greens.

Kohlrabi Roots

Kohlrabi roots that are misshapen can be placed in a sand-filled box, described in the section on cellar gardens, to provide a winter-long supply of salad greens. Kohlrabi itself stores well enough and does not need to be winter grown.

Lemon Balm (Melissa)

Lemon balm can be grown from seed or propagated through root division and cuttings. Plant seeds in a loose, fine soil in a seedpan. When the seedlings are one inch high, thin them out to two inches apart. When they reach four inches transplant them into a pot filled with sterile potting soil. Keep the soil moist throughout the growing.

If lemon balm is grown indoors during the winter, place it under fluorescent lights that are turned on 14 hours a day and positioned five inches above the plant tops. Water the plant well to prevent leaves from yellowing.

Harvest lemon balm by cutting off the entire plant two inches above the soil level. If the plant is sufficiently watered and fertilized it will produce several more cuttings each season. The cuttings should be dried within two days of the harvest. For use in a tea, both leaves and stems should be dried.

Lemon Verbena

Lemon Verbena is a deciduous shrub that is difficult to propagate and tends to lose its leaves when brought indoors at summer's end before the first fall frost. The leaves are used mainly for teas, although they can also be used as a substitute for lemon or mint in stuffings, poultry and fish dishes.

Since seeds do not germinate easily and cuttings are difficult to root, it is best to purchase the plant from a nursery or garden shop. If the plant has grown outdoors all summer, pot it for indoor growing and allow it to acclimate itself to the pot outdoors. Bring it inside before the first frost. Fertilize the plant at least once every two weeks indoors and mist the leaves once a week with lukewarm water.

The plant does best in a well-drained soil consisting of a mixture of loam, sand, leaf mold or humus and dried cow manure. A bit of bone meal in the mixture is also helpful. It can be grown under lights or in a sunny window.

Lemons

Lemons can be grown indoors or in a greenhouse. Buy a dwarf variety from a nursery and plant in the same manner as the grapefruit. Follow the same methods of cultivation as those described for grapefruit.

Lettuce

Lettuce can be grown indoors under lights, in the greenhouse and in a cold frame. Some gardeners claim that lettuce will grow outdoors with the help of a cold frame 300 days out of the year.

The best varieties for indoor growing are Summer Bibb, Tom Thumb, Buttercrunch, and Grand Rapids. Lettuce requires high humidity to germinate and grow. Use trays with one or two inches of pebbles or rock chips and add water to just under the top of the rocks. Place both germinating mediums and pots on the pebbles. The container bottoms should not touch water.

To germinate lettuce, sow the seeds in Jiffy 7 pellets. Place in the pebble containers and position under a fluorescent light turned on 14 hours a day. The light should be at least five inches from the top of the Jiffy 7 pellets. Leaf and bibb-type lettuce can be germinated in a cool 65°F area, but head lettuce prefers more soil heat. Leaf types of lettuce do not need to be thinned and can be placed directly in the growing container. Head lettuce plants should be placed one to a pot.

Since lettuce grows best in a rich, moist soil, add 1/4 cup of dried cow manure or bone meal to each quart of sterile potting soil. Be sure to test the pH of the soil and adjust to bring it to neutral or slightly acid. Bone meal or manure increases alkalinity and if the soil used is already alkaline the pH will have to be adjusted back to neutral or slightly acid. Water frequently and fertilize every ten days with a good liquid fertilizer. A top dressing of nitrate of soda is also useful. Keep the atmosphere cool and damp.

Seed lettuce in the same manner for greenhouse cultivation. For greater space efficiency, however, transplant the lettuce from peat pellets to a growing bed when the plants are two to three inches tall. Bibb varieties usually require an eight-inch square while head varieties need a square foot each. The same space requirements hold for cold frame cultivation, but the seeds can be sown directly into the ground and thinned as they appear. The thinnings can be used for table salad greens.

Limes

Limes can be grown indoors or in a greenhouse. Buy a dwarf variety from a nursery and plant in the same manner as the grapefruit. Follow the same methods of cultivation as those described for grapefruit.

Marjoram

To grow marjoram indoors, plant the seeds in Jiffy 7 pellets and place under fluorescent lights for at least 14 hours a day. The lights should be positioned five inches above the tops of the pellets. Depending on the temperature and moisture, marjoram will take anywhere from eight days to four months to germinate.

When roots show through the sides of the pellet, the plants are ready to be transferred to a three-inch pot. The pot should be filled with good draining, sandy soil that is somewhat alkaline. To each quart of sterile potting soil, add one tablespoon of limestone and 1/2 cup of coarse sand.

Grow marjoram under fluorescent lights that are turned on 14 to 18 hours a day. The amount of time should be determined by how well the plant grows. Do not fertilize.

Two harvests of marjoram are recommended. The plant is ready for the first harvest when green, ball-shaped tips appear at the end of the stems. Cut off all of the plant at a level one inch above the soil. This will stimulate a second and much fuller growth. When the ball-shaped tips appear a second time, cut the plant back again.

Mint

Mint is an excellent plant to grow indoors under lights or in a window. There are many varieties of mint—spearmint, peppermint, applemint, orangemint, watermint, etc.—each with an individual flavor of its own. But the growing conditions for all mints are approximately the same.

Mint is best grown from cuttings rooted in water and transplanted to pots of sterile soil that is not too rich. Water well since the roots need to be wet.

Mushrooms

The four essentials for successful mushroom culture are darkness, manure, temperature and moisture. Darkness is easiest to achieve in a cellar or underground area where the other three essentials can be maintained.

The manure must be fresh, hot horse manure with plenty of *straw*. Shavings, sphagnum moss or any other kinds of bedding are not acceptable nor is disinfected manure. The pile of manure should be wetted down and turned after a few days. This process should be repeated and continued for about 30 days. By this time the temperature in the center of the pile will fall to about 70°F to 75°F.

Spread layers of the 70°F to 75°F manure six to eight inches deep in beds or trays and wait until the temperature drops to between 58°F and 65°F. The air should also be maintained at this same temperature. A fluctuation of temperature can be fatal to the mushrooms.

Buy spawn from a nursery and plant in the prepared bed in two inch blocks about ten inches apart and no more than two inches deep. Little or no water should be added at this time. Add *tepid* water only if the manure shows a tendency to dry out. After about three weeks the entire bed should be well covered with tiny, white, thread-like rootings. At this time, spread a layer of garden loam about one inch deep over the entire bed. The mushrooms should appear after six to seven weeks and should bear for two to three months. The manure should then be disposed of in a compost pile but not used for mushrooms again.

During the growing period the temperature and moisture are extremely important. Air temperature should be as near as possible to 56°F to 58°F. The beds should be moist but never soaked and only tepid water used. The humidity should be maintained at 70% to 75%. Harvest the mushrooms at the size preferred. Clean the area thoroughly after the bed is spent, disinfect and allow it to rest a few weeks before starting a new bed.

Mustard Greens

Mustard greens can be grown in a cold frame. Start the seeds directly in the soil of the frame and thin to three inches between plants, using the thinned greens for salads. Mustard greens grow best in a soil containing well-rotted manure or a good commercial fertilizer. The plants may also be lightly top dressed with nitrate of soda. They like neither heat nor dryness, and adequate moisture is necessary for their cultivation.

Harvest the greens by picking the outer leaves. Successive plantings will yield more crops.

Mustard greens can be grown under lights using the same methods as those described for lettuce.

Nasturtium

Nasturtiums should be seeded two or three to a pot and grown under fluorescent lights positioned at least five inches above the pot and the subsequent growth. Since the plants like well drained soil, add 1/2 cup of sand to each quart of sterile potting soil. The soil should not be too rich or the plants will be all foliage with few flowers.

The leaves and flowers, which can be harvested at any time, are edible, as are the seeds. The leaves and flowers can be used fresh in salads or sandwiches in place of, or in addition to, lettuce. The seeds are picked when the clusters are half grown. Clean the clusters and cover with boiled cider vinegar. Tightly seal the jar and store in a cool place. Eat the pickled seeds as a snack, as the Chinese do.

Onions

Onions can be grown indoors under lights, in a cold frame, in the greenhouse, or, if well mulched, outdoors. While onions can be grown from seed, sets—little clumps of onion bulbs—will produce better and grow faster. Indoors, for example, sets will produce green onions within three weeks.

For indoor cultivation, place onion sets in a good-sized container, a bulb pan or another container that will accomodate the size onion desired. Fill the container with sterile potting soil, adding 1/2 cup of sand and 1/4 cup of bone meal for each quart of soil. Do not overwater or the onions may rot. Feed with liquid fertilizer once every ten days.

For cold frame growing, plant the sets an inch deep and an inch apart in rows that are six inches apart. The onions should be planted in early fall to give them a chance to develop before the first freeze. In some climates onions can be planted outdoors at the same time and protected throughout the winter with a heavy mulch of six or more inches of straw. They will continue to produce throughout the winter. Sets should be planted in a greenhouse in the same manner as in a cold frame.

The time to harvest onions is a matter of taste. Early harvesting will yield plants the size of a green onion. The longer the plants are left in the earth the larger they will grow. Outdoors, however, when the green tops shrivel and fall over, the onion has grown to its full size and should be harvested rapidly.

Oregano

Oregano can be grown and used fresh in a variety of dishes, particularly Italian specialties.

Root a cutting of oregano in water or a propagating medium, or sow seeds in Jiffy 7 pellets, three seeds to a pellet since the plant prefers to grow in clumps. Place the pellets under fluorescent lights that are turned on 14 hours a day and positioned five inches above the top of the pellets. When roots show through the pellets, place the pellets into three-inch pots filled with sterile potting soil, adding 1/2 cup of sand and one tablespoon of limestone for each quart of soil. The plants can be grown to maturity either under the lights or on a window sill with direct sunlight.

Pinch the tops to make a compact, bushy plant, the leaves of which contain more essential aromatic oils and are more flavorful than those of spindly plants. Oregano is ready to harvest, if it has not been continuously picked, when flowers appear. In any event, six weeks after planting cut off all stems one inch above the growing center. This will stimulate a more dense, bushy plant.

Parsley

The Greeks used to weave wreaths of parsley with which to crown victors of athletic events. Parsley itself is one of the prizewinners in a winter garden because it can be cultivated under so many different conditions.

Parsley roots can be planted in a sand-filled box, as described in the section on cellar gardens, for a winter-long supply of this nutritious garnish. Parsley can also be planted in cold frames and grown in greenhouses. Greenhouse parsley prefers a nighttime temperature of 55°F.

For indoor gardening under lights, parsley is germinated in Jiffy 7 pellets and then placed in a three-inch pot of sterile potting soil. Since parsley takes a while to germinate under normal circumstances, the process can be speeded by soaking the seeds overnight in a glass of lukewarm water. Soaking breaks down the protective covering on the seeds and they will germinate rapidly in two weeks instead of the normal six weeks. This treatment also works for parsley seed grown under other conditions.

Indoors, place the propagating pellets in a plastic tray containing a layer of pebbles and water. The pellets should rest on the pebbles, but not come into direct contact with the water. After the seeds germinate, transfer the pellets to pots filled with a sterile potting soil that is not too rich and continue to grow the plant under the fluorescent lights. The light should be turned on the plants for 14 hours per day and should be positioned five inches above them. Since parsley likes to be kept moist, daily misting helps. Feed with liquid fertilizer once every ten days.

To harvest, take cuttings from the outside part of the plant since parsley grows from the center.

Parsnips

Parsnips can be grown in a cold frame for the roots and in the cellar for the greens.

Seed parsnips in the cold frame in rows three inches apart, with about one inch between the seeds. Do not cover with too much soil since the parsnip seed cannot push up through a heavy soil cover. A few radish seeds mixed in with the parsnips will help them break through the top of the soil. When parsnips grow to a height of two inches, carefully thin the seedlings to a distance of three inches apart and place the delicate thinned seedlings in another part of the cold frame. Parsnips are extremely tolerant of cold weather, although they like a good mulch. They grow best in a soil that is deep, rich and sandy. Some gardeners insist that cold weather improves the parsnip's flavor by changing starch to sugar.

To cultivate a winter supply of parsnip greens, place misshapen parsnip roots in a sand-filled box as described in the section on cellar gardening.

Peas

Since peas like cool temperatures they are a natural for the cold frame. Miniature varieties such as Wando, Mighty Midget, Tiny Tim and Dwarf Gray Sugar are best for the cold frame, although edible pod varieties also grow well. Peas need a trellis on which to climb so the back portion of the cold frame is the best location for the plants since it gives them room to climb.

To plant the seeds, dig a trench one inch deep. Water the soil until it is saturated. Then place the pea seeds one inch apart and cover with soil that contains a good amount of well-rotted manure. If another row is desired for the other side of the trellis, plant it three inches away from the first. Peas like cool temperatures and plenty of water. Harvest the pods before they get too fat; peas tend to become tough and inedible not very long after maturity.

Peas can also be grown indoors under lights, provided they have space to grow, means to climb, and conditions approximating those described above.

Peppers

Peppers are a challenging winter crop because they require very warm temperatures and a long growing season. For this reason they should only be attempted in a warm greenhouse or a hothouse.

If the greenhouse temperature is at least 75°F the peppers can be seeded in the greenhouse. If not, start them indoors in Jiffy 7 pellets placed in the hottest part of the house to insure germination.

After germination, transplant the seedlings into good-sized pots filled with a well-drained sterile soil that contains a good amount of well-rotted manure or commercial fertilizer. Water frequently and feed with a water-soluble fertilizer about every ten days. Pinch off the top leaves to force the plants into a bushy growth. Stake the plants if necessary.

In harvesting peppers, leave at least a half inch of stem. This protects and stimulates the plant and helps the peppers keep better for storage.

Pineapple

Pineapples can be cultivated indoors, but don't expect to create a Dole plantation in your home.

Buy a pineapple at the supermarket—one with a fresh-looking, green, springy bunch of leaves. Cut off the top along with an inch of the fruit and put it in a small amount of water away from any direct sunlight. After four or five weeks it will root. Plant it in a pot filled with sterile potting soil to which is added 1/2 cup of sand per quart of soil (pineapples prefer porous, thoroughly drained soils). Place the plant in direct sunlight and spray the crown. Spray with a liquid fertilizer diluted to one-fourth the recommended strength.

Once the plant is acclimated to the pot, the object is to force the plant to flower, since the flowers precede the fruit. Don't rush the process, however. Plants grown from crown cuttings can take three years before they are mature enough to bear fruit.

Place the plant in a plastic bag along with an apple. The apple emits ethylene gas which, trapped in the plastic bag, stimulates the plant to produce a cluster of 100 or more tiny flowers that will each turn into one segment of the fruit. The plastic bag should be left on for five days and then removed. During this time the temperature must be maintained between 75°F and 85°F. When new growth starts coming up, support the plant with a stake. Cut off the pineapple when ripe and treat family and friends to this homegrown rarity.

Radishes

Radishes can be grown in a cold frame, greenhouse, or indoors under light. For both cold frame and greenhouse planting, sow the seeds in rows four to six inches apart and covered with not more than 1/2 inch of soil. After germination, the seedlings should be thinned to about one per inch, otherwise the crop will be greens only. Keep the plants cool and well watered. Harvest when the plants reach maturity, usually three to four weeks after planting.

For cultivation indoors under lights, plant the seeds in pots big enough to accommodate the roots. Use sterile potting soil and add 1/2 cup of sand to each quart of soil. Water well, keeping the plants cool and on the moist side or the radishes will be cracked and dried out. Give the plants 14 hours of light each day and position the light five inches above the plant tops.

Rosemary

While rosemary can be grown from seed, the low rate of germination suggests that it is better to propagate the plant from cuttings. To root a cutting, submerge four inches of a six inch cutting in well-soaked vermiculite and cover with a plastic bag. Water regularly; once the plant dries out it cannot be salvaged.

When well-rooted, transplant the cutting into a pot filled with sterile potting soil to which has been added 1/2 cup of sand and a tablespoon of limestone for each quart of soil. Place the plant where it will receive reflected light with some direct sunlight. Again, never allow the plant to dry out. Pinch off the tops to promote a compact bushy plant, the leaves of which are more flavorful than those from spindly plants.

Rhubarb

Rhubarb is the easiest vegetable to force in the cellar. Leave the clumps in the ground until they freeze hard. If local winter weather does not bring a hard frost, put rhubarb roots in a freezer where the temperature is between 10°F and 32°F. Leave them in the freezer for two weeks and then force them.

Place the rhubarb roots in a box of soil with cow manure, horse manure, peat moss or well-rotted compost under each root clump. Begin forcing the roots in January. Place the box of earth in a dark place, or cover it to keep all light out since rhubarb gives better stalks in a dark, moist place. Do not eat the leaves; they may be toxic.

The ideal temperature for forcing rhubarb is between 50°F and 60°F. The stalks will come up faster at temperatures over 60°F, but these seem to be stalks of inferior color and quality. Any temperature below 50°F slows down the growth.

Rhubarb can also be forced outdoors in a cold frame.

Cover rhubarb left in the ground with a portable cold frame and cover the cold frame with a blanket to keep out all light.

Rutabaga

Rutabaga roots can be placed in a sand-filled box, described in the section on cellar gardening, and harvested for salad greens during the winter.

Rutabaga can also be grown in a cold frame, but is better stored for the winter from summer crops since it stores so easily. For cold frame cultivation, plant two or three seeds to the inch and thin seedlings to about three inches apart. Rows should be four or five inches apart.

Sage

Sage can be propagated from seeds or from cuttings rooted in vermiculite and sand. To pot sage, add 1/2 cup of sand and one tablespoon of limestone to each quart of sterile potting soil and plant one seed to a pot. Keep sage on the dry side and place the plant under fluorescent lights turned on 14 hours a day or on a window sill that receives direct sunlight.

Leaves can be harvested when they are mature. In the first year, harvest only leaves and stems high up on the plant. During the second year, at least two harvests are possible. At least once a year trim off any woody growth on the stems. One plant will not last much more than three or four years since the stems will become very woody, tough and unusable.

Salsify

Salsify (also known as vegetable oyster or oyster plant because of the delicious flavor of the cooked root that is faintly reminescent of oyster) can be grown for its roots in a cold frame or used in a sand-filled box as described in the section on cellar gardening to grow a winter-long supply of salad greens.

To obtain the best growth, salsify should be seeded as soon as possible after the last spring frost. Salsify has a long growing time to maturity and the plant should be nearly mature before the first killing frost of fall arrives. The best soil for salsify is a rich loam to which manure has been added. Salsify will not grow well in a tight, heavy soil with too much clay or silt. Manure used should be well-rotted and fine. Fresh, coarse manure makes the roots grow rough.

Seed salsify in rows 12 to 18 inches apart, two or three seeds to the inch, and cover with 1/2 inch of soil. Later, thin the seedlings to two to four inches apart. The plant roots can be partially harvested in the fall and stored or used to grow greens, with the remainder dug up during the winter or in the spring. Mulch the roots left in a cold frame during the winter. Salsify is a hardy root and gardeners say its flavor is improved by frosts.

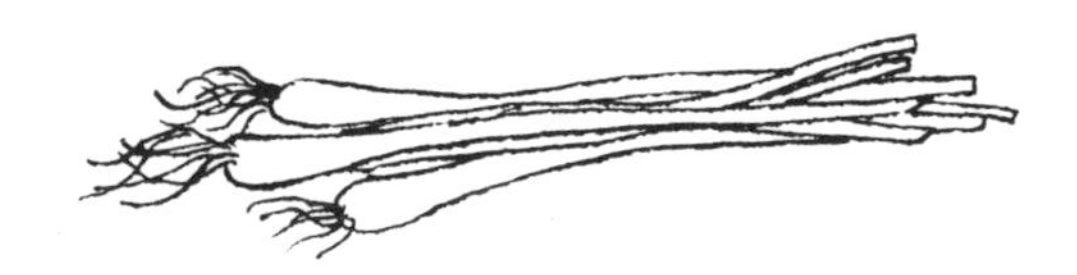

Savory

While both summer and winter varieties of savory have been grown indoors under lights, most gardeners favor summer savory because it is less woody than the winter variety.

Sow three or four seeds of summer savory in each Jiffy 7 pellet or peat pot filled with sterile potting soil. Place under fluorescent lights that are five inches above the top of the propagating container and turned on at least 14 hours a day. When roots show through the sides of the Jiffy 7 pellet, plant it in a pot of sterile potting soil, adding 1/2 cup or sand per quart of soil. Keep moist because summer savory needs plenty of water to grow to its maximum height. Mist at least twice a week for proper humidity. The nutrients contained in the soil should be sufficient because summer savory is a modest feeder.

Harvest summer savory as soon as the plant reaches a height of six inches. Cut only the tops of the plants which will prevent the plant from flowering. If summer savory does flower, cut the whole plant as soon as the flowers open or the leaves will turn yellow, then brown and finally curl up.

Winter savory is planted the same way, except that it prefers a coarse soil that is not very nutritious. For winter savory, add 1/2 cup of gravel to each quart of potting soil instead of the sand added to the soil for summer savory. Do not water winter savory as much as summer savory; too much water will decrease winter savory's hardiness.

Shallots

In winter, shallots are best grown indoors under lights. Unlike onions, they are not hardy enough to take the outdoor cold in most climates, even in a protected cold frame.

Plant shallot sets in a good-sized container—a bulb pan or another container large enough to accomodate the full-grown bulbs. Fill the container with sterile potting soil to which 1/2 cup of sand and 1/4 cup of bone meal have been added for each quart of soil. Do not overwater or the shallots may rot. Feed with liquid fertilizer once every ten days. Give the shallots 14 hours of light each day, with the lights positioned five inches above the tops of the growing plants.

After harvesting shallots, condition and store them in the same manner as onions or garlic.

Spinach

Spinach is a good cold frame crop, since it prefers cool weather and does not fare well in heat. Sow spinach seeds outdoors in September or October, before the first hard frost, in rows 12 to 18 inches apart. Place seeds two or three to the inch and later thin to one to three inches between plants. Spinach can also be seeded in Jiffy 7 pellets indoors and later transfered to the cold frame.

Spinach prefers a neutral to slightly alkaline soil. If the cold frame soil is acid, add lime at the necessary rate to bring the pH to between six and eight. Water well and fertilize often with a good commercial fertilizer that contains an abundance of nitrogen. A top dressing of nitrate of soda is also useful.

Harvest spinach by taking the entire plant. Make successive seedings to have a continuous supply.

Tarragon

The tarragon favored for its aroma and flavor is commonly called French tarragon and does not seed well. Another type of tarragon, Russian, seeds copiously but does not have the same aroma and flavor of French tarragon.

Tarragon must therefore be propagated by cuttings or root division. Once a tarragon cutting has been rooted, place it in a pot with sterile potting soil to which 1/2 cup of sand or fine gravel and one tablespoon of limestone has been added for each quart of soil. The root system of tarragon spreads horizontally rather than vertically and a full grown tarragon plant will require a ten-inch pot to provide enough room for the roots. To achieve and maintain full growth—up to 12 to 18 inches indoors—place the tarragon plant under fluorescent lights that are turned on at least 14 hours a day.

Do not overwater tarragon; let the soil dry out for at least a day or two between waterings. Spray the leaves with a mister once a month and the aroma of new mown hay will permeate the house. Tarragon also benefits from feedings with a liquid fertilizer once every two weeks.

Harvest tarragon as needed. If the lower leaves start to turn yellow, either a sign of aging or inadequate fertilization, cut the whole plant to within two or three inches of the soil. Once yellowing begins it will progress and cannot be reversed. Dry the cut leaves and store for later use.

Thyme

The last member of the parsley, sage, rosemary and thyme quartet should be seeded in Jiffy 7 pellets and placed under fluorescent lights that are turned on 14 hours a day and positioned five inches above the pellets. When the thyme seedlings have grown large enough, transplant them into pots filled with sterile potting soil to which 1/2 cup of sand and one tablespoon of limestone is added for each quart of soil.

Water sparingly because thyme likes to be dry. Harvest thyme just before flowers begin to open, the point at which the plant leaves have the largest amount of essential oils. If the flowers are allowed to open the fragrance and flavor are diminished. Cut the entire plant off to a level about two inches above the soil. Harvest subsequent growths in the same way. Thyme can be used fresh or dried.

Tomatoes

Growing tomatoes indoors under lights or in the greenhouse requires close attention to the plants and maintenance of the proper temperatures during germination and pollination. For indoor growth, choose one of the midget varieties such as Red Cherry, Red Pear, Yellow Pear, Yellow Plum, Small Fry, Tiny Tim Red, Tiny Tim Yellow, Patio, Presto, Pixie, Tom Thumb or Early Salad. The smaller plants generally produce smaller tomatoes, but space indoors is at a premium and these smaller varieties will do better indoors than the larger ones.

Seed Jiffy 7 pellets with one tomato seed each and place the pellets under fluorescent lights turned on 14 hours a day and positioned five inches above the container. To prevent seeds from rotting and to promote successful germination the temperature should be between 75°F and 80°F. Water the seeds thoroughly and do not let them dry out. Once seedlings appear, continue to water thoroughly. If tomatoes dry out they will die. When the true leaves appear and the plants are at least two inches tall, place them in pots filled with sterile potting soil, one plant to a pot. Fertilize the plants every ten days.

Getting the plants to pollinate may be a little tricky indoors because the wind does not jostle the plants. Temperature is also important. There is some disagreement among gardeners as to whether temperature or motion is more important. One school of thought holds that the correct temperatures stimulate the blossoms to grow at the proper rate to allow self-pollination without any need for motion or the gardener's help. The best temperature for obtaining a good fruit set is between 80°F and 90°F during the day and between 65°F and 75°F at night. Tomatoes can also be pollinated by gently shaking the plant so that the pollen lands on the stigma, or by using a fine paintbrush to actually transfer the pollen to the stigma of the blossom.

Tomatoes grown in the greenhouse should be seeded in the last two weeks of June for a fall crop and the last two weeks of December for a spring crop. This is much better than trying to grow one crop during the winter when the plants have a difficult time setting and developing fruit because of the fewer hours of sunlight.

Sow seeds in a bed. When they are an inch tall transplant to three or four-inch peat pots; when the plants reach three or four inches in height transplant to the growing beds. Plants should be far enough apart to provide four or five square feet of growing space for each plant. Fertilize every ten days to two weeks with a liquid fertilizer, mixing an ounce of fertilizer in a gallon of water—enough to feed four plants.

Successful greenhouse tomatoes require a minimum growing temperature of 60°F, adequate ventilation, and a relative humidity below 90 percent. Water the soil directly and don't let any of the water touch the plant leaves. Use fungicides if necessary to prevent pests.

Turnips

Turnips are cool season crops. They can be planted as late as September in a cold frame and will mature within a month and a half. The best time to seed, however, is late July or early August.

Sow turnip seeds in rows that are 12 to 15 inches apart and thinned to between two to five inches depending on the size turnip produced by the variety sown. Turnips grow best in a loose, friable, rich soil. An acid soil should be treated with lime. Feed with any good commercial fertilizer.

Harvest turnips before the first light frost and before the roots become too large, or the turnips will be woody and bitter. Turnips can also be used in a sand-filled box, described in the section on cellar gardening, to produce a winter-long supply of salad greens.

Watercress

To grow watercress indoors, plant the seeds in a bed of potting soil to which one tablespoon of limestone is added for each quart of soil. Place the container on a partially lit, cool window sill. The key to growing watercress successfully in the house is to keep it wet at all times. Seedlings will mature in a little less than two months.

Watercress will also root in wet sand or a sandy soil. The seedlings can be placed in a cold frame to mature, provided they are shaded, kept constantly wet, and the outdoor temperature is not too severe.

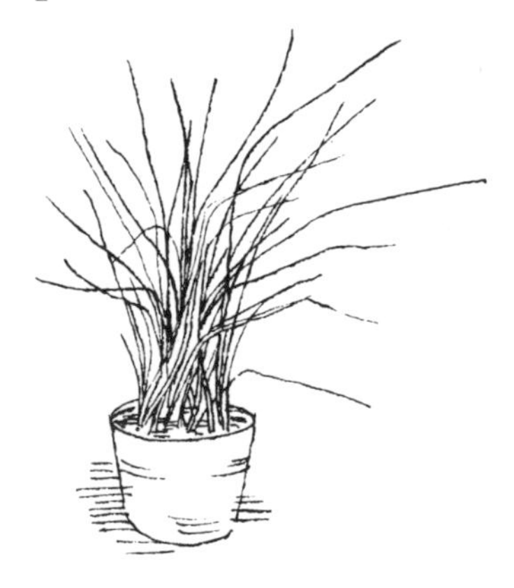

INDEX

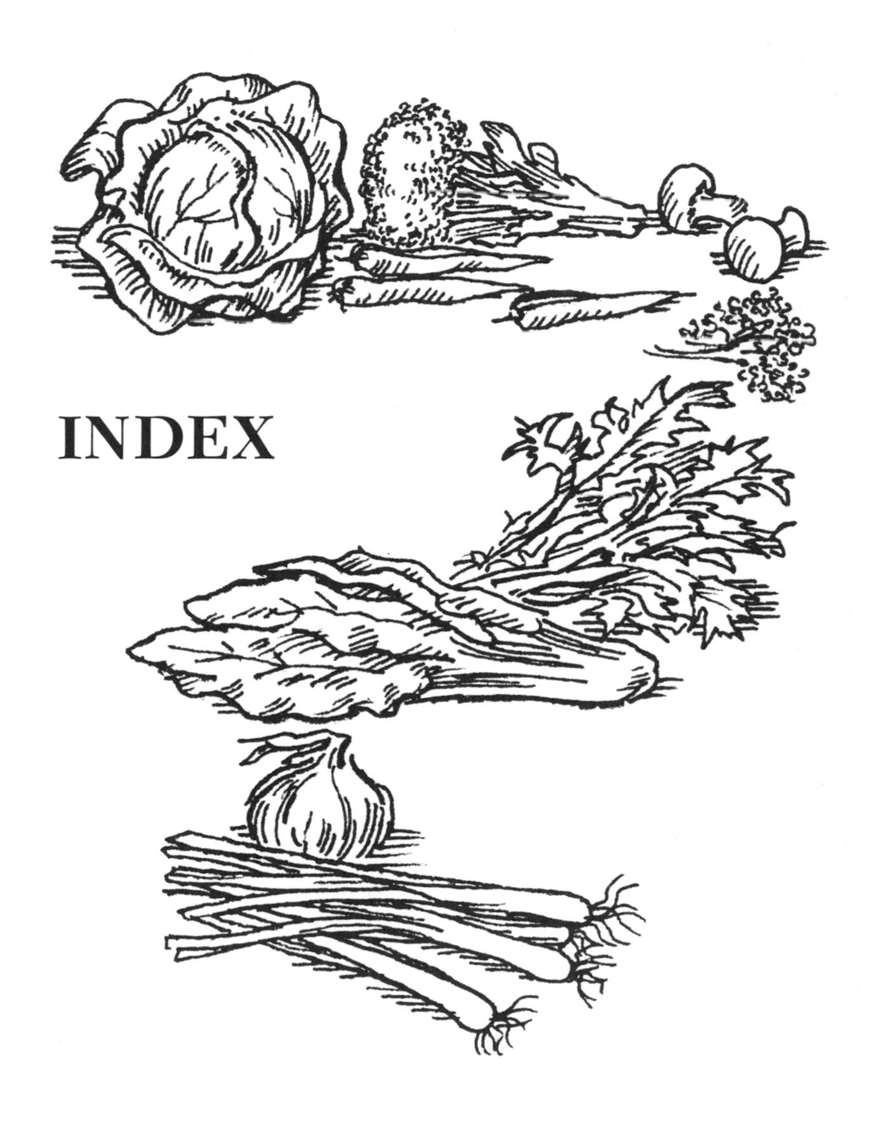

This book is set in a typeface called Caledonia and was composed through the Linofilm process by P&M Typesetting, Inc., Waterbury, Connecticut. The book was printed by The Murray Printing Company, Forge Village, Massachusetts, and bound by The Book Press, Brattleboro, Vermont. The book was designed by Miller-Hormel Studio.